MATH REVIEW TOOLKIT

GARY L. LONG
Virginia Tech

SHARON D. LONG

Introductory Chemistry
SECOND EDITION

Nivaldo J. Tro

PEARSON

Prentice
Hall

Upper Saddle River,

S0-DVE-947

Project Manager: Kristen Kaiser
Senior Editor: Kent Porter-Hamann
Editor-in-Chief, Science: John Challice
Executive Managing Editor: Kathleen Schiaparelli
Assistant Managing Editor: Becca Richter
Production Editor: Rhonda Aversa
Supplement Cover Manager: Paul Gourhan
Supplement Cover Designer: Joanne Alexandris
Manufacturing Buyer: Alan Fischer

© 2006 Pearson Education, Inc.
Pearson Prentice Hall
Pearson Education, Inc.
Upper Saddle River, NJ 07458

The author and publisher of this book have used their best efforts in preparing this book. These efforts include the development, research, and testing of the theories and programs to determine their effectiveness. The author and publisher make no warranty of any kind, expressed or implied, with regard to these programs or the documentation contained in this book. The author and publisher shall not be liable in any event for incidental or consequential damages in connection with, or arising out of, the furnishing, performance, or use of these programs.

Printed in the United States of America

10 9 8 7 6 5 4 3 2 1

ISBN 0-13-147066-3

Pearson Education Ltd., *London*
Pearson Education Australia Pty. Ltd., *Sydney*
Pearson Education Singapore, Pte. Ltd.
Pearson Education North Asia Ltd., *Hong Kong*
Pearson Education Canada, Inc., *Toronto*
Pearson Educación de Mexico, S.A. de C.V.
Pearson Education—Japan, *Tokyo*
Pearson Education Malaysia, Pte. Ltd.

Contents

Acknowledgement

The authors (GLL and SDL) would like to acknowledge the contributions of Doris L. Lewis on the Sections 12 and 13 of this Toolkit on Chemistry and Writing, and Careers in Chemistry. Also acknowledged are the contributions of Timothy Smith and Diane Vukovich on Section 14 concerning Study Aids for Chemistry.

A successful study of general chemistry requires you to use your skills of memorization, logic, and mathematics. From my years in the classroom, I have found that students who are unsure about their math skills generally do poorly on their exams. To put this statement into context, imagine that you are holding a chemistry exam containing 20 questions, with 15 of them involving the use of mathematical equations. If you have trouble with the first few math questions and have only an hour to complete the exam, any weakness in math skills will limit your performance on the exam, regardless of how many facts and figures you have memorized.

The math skills that a student needs to successfully complete general chemistry are basically what you learned in high school: algebra and trigonometry. It is the purpose of this booklet to guide you chapter by chapter in the mathematics that are used in the second edition text of **Introductory Chemistry**. The math skills for each chapter are outlined and examples are worked in this booklet.

The case of a life sciences student planted the idea for this booklet. This young man was going around for the seventh time in general chemistry. He could never manage to pass the first test, so he would drop out and wait until the next semester to try again. Fortunately, he came to me at the beginning of the course. He was desperate; he could not graduate until he passed chemistry. After speaking with him, I found his limitation was not inadequate math training, but a basic fear of math problems. He would stumble over the math problems on the exam. Over the course of the semester I met with the student and explained how to perform these calculations using the methods that are described in this booklet. The student successfully completed the course with a B average.

Although we cannot absolutely guarantee your success in chemistry with the use of this booklet, mastery of the material presented here will enhance your ability to work problems in general chemistry. Take a few moments and complete the "Self Test for Math Skills" at the back of this book. It will help you assess your skills. Use this guide to help you reinforce your math skills.

Chemistry is an exciting science that touches every aspect of our lives. It is our hope that this toolkit will demystify the mathematics used in general chemistry and help you discover this excitement.

Gary L. Long
long@vt.edu

I will never forget my own struggle with chemistry at Wake Forest University. As a freshman, I had not yet developed productive study habits. I faced every new chemistry chapter with a certain dread. There is a tendency to procrastinate taking hold of something we fear or find distasteful, or so it was in my case. However, if you cope with chemistry in this manner, as I did, the stress of catching up and preparing for a test will cost you your well being until the test is over. The two grades of C that I made in freshman chemistry taught me this painful lesson. The point being, even if you dislike chemistry and are taking it only to satisfy the requirements for your major, make your effort consistent.

Many college students have been forced to change their career dreams because they could not make it through chemistry. This unpleasantness does not give this science a good name nor does it give chemistry professors an easy grace at social functions when asked their occupation. Some of you may need to change your career option to a different discipline for which you are gifted. A very kind English professor pointed this out to me at a time when organic chemistry was causing me great distress. Time is too valuable to spend it in the wrong place. However, for those of you heading in the right direction, we do not want chemistry to be a stumbling block for your dreams.

We hope that your experience with chemistry will be a victorious one and that in some small way this math book will make a difference. Write us and tell us your stories as well as your triumphs.

Sharon D. Long
sdlong@vt.edu

SECTION 1

Harnessing the Power of Your Calculator

The best friend you have in a chemistry class is your calculator. If used correctly, it will save you precious seconds during a test, make doing homework problems outside class much more efficient, and allow you to spend more time on the concepts rather than laboring over the calculations.

In your general chemistry course, you will encounter problems using arithmetic operations (addition, subtraction, multiplication, and division), trigonometric operations (cos, sin, $\cos^{-1}$, and $\sin^{-1}$), and logarithmic operations (log, ln, 10^x, e^x, powers, and roots). You also will deal with scientific notation in many of these problems. While your textbook has an excellent appendix on mathematics for general chemistry, we present in this toolkit methods for using your calculator to solve these calculations.

Take the time to acquaint yourself with the capabilities of your calculator. Please try the practice problems in Section 1.12 of this toolkit. You will find your calculator to be a powerful tool that you can harness for your study of general chemistry.

1.1 Ten-Dollar Calculators That Work

Many students have difficulty in the first few weeks of the semester because they don't have the type of calculator they need in order to quickly and efficiently perform the tasks the instructor and textbook will require of them. You may sit down in your first class with a four-function calculator that just does the basic operations of adding, subtracting, multiplying, and dividing (the type the bank gives you when you open a checking account). This type of calculator is not powerful enough to do many of your chemistry problems. On the other hand, you don't want to purchase an engineering calculator that has so many functions it is intimidating to look at and has a manual as thick as *The Hobbit!* You need a scientific calculator that will help you quickly and easily dispatch the problems set before you in your class, especially on tests. The calculator should have the following capabilities:

- The four basic arithmetic functions;
- A button to enter scientific notation (EE or EXP on most calculators);
- Buttons to easily enter logarithms.
- Buttons to convert between scientific notation and decimal form;
- Buttons to do statistical calculations (you won't need them in this class, but many of you will probably take statistics later in your program);
- At least 10 places on the display so you can display fairly large numbers in decimal form, (called 10 + 2 on some models).

Below is a list of calculators that are inexpensive (around $10) and do the job you need a calculator to do:

- Casio FX-250HCS
- Texas Instruments TI-30Xa
- Sharp EL-501

1.2 The First Five Buttons—The Basic Operations

Nothing is easier than multiplying or dividing on a calculator. If you want to multiply 2.71 × 1.91, you simply punch

2.71 [×] 1.91 [=]   The display reads 5. 1 761

When you divide two numbers, the process is the same. Let's divide 3.75 by 1.2. This problem can be presented to you as 3.75 [÷] 1.2

or 3.75/1.2 or $\dfrac{3.75}{1.2}$. No matter how the division is written, it is entered into the calculator as

> 3.75 [÷] 1.2 [=]   The display reads 3. 1 25

Most of the time in chemistry you will do several operations in one problem. An example of this type of problem is

$$2.875 \times \frac{1}{3.125} =$$

Many students enter the operations like this:

> 2.875 [×] 1 [÷] 3.125 [=]   The display reads 0. 92

It is much simpler and less prone to error if the multiplication step is left out when all you are doing is multiplying by one. You simply enter

> 2:875 [÷] 3.125 [=]   The display reads 0. 92

Later, when you do conversions, you will do many multiplications and divisions in one problem. Let's try one.

$$\frac{2.375 \times 4.15}{0.03125 \times 1.577} =$$

Many students do the problem like this:

First, they multiply the numbers in the numerator

> 2.375 [×] 4.15 [=]   The display reads 9. 85625

Then they write this answer down, push [C] to clear the display, and then multiply the numbers in the denominator:

> 0.03125 [×] 1.577 [=]

The display reads 0. 04928125

Then they write this number down, clear the display, reenter 9.85625, and finally divide by 0.04928125 to get the answer 200.

In other words, they multiply the numbers in the numerator first, erase the answer, then multiply the numbers in the denominator, reenter the first answer, and then do the division.

There is an easier way to solve this equation. Do the multiplications and divisions in order from left to right. Press [×] before numbers that are in a numerator (after the first number) and press [÷] before any number in a denominator.

2.375 [×] 4.15 [÷] 0.03125 [÷] 1.577 [=]
The display reads 200

You can change the order, as long as you press [×] before a number in the numerator and press [÷] before a number in the denominator.

2.375 [÷] 0.03125 [÷] 1.577 [×] 4.15 [=]
The display reads 200

Let's try another.

$$1.44 \times \frac{1}{16} \times \frac{13.3}{0.20} =$$

We punch

1.44 [÷] 1.6 [×] 13.3 [÷] 0.20 [=]
The display reads 5. 985

Doing problems this way can save you much time on tests.

1.3   Rounding Off and Significant Figures

If you divide 2 by 7 on your calculator, the display reads 0.285714286. Ten decimal places can be quite cumbersome. Fortunately, you do not have to write all those numbers when you answer a problem; you can round off your answer. Answers beyond a certain number of places are not just cumbersome, they are misleading. So we round our answers off to a reasonable number of decimal places.

In order to round off, we have to understand significant figures (sometimes referred to as sig figs). In chemistry (and in all science), we do not work with pure numbers as in algebra class. We work with measurements. If you take your body temperature with a fever thermometer, it may look like the figure below. The mercury stops between 98 and 99 degrees. It looks like it is between 98.2 and 98.4 degrees (each small line on the thermometer is 0.2 of 1 degree), so we might guess 98.3. In all measurements, the last digit is a good guess and your guess might be a little different from mine. This temperature (98.3 degrees) has three significant figures. If we round it off to two places, it will have two significant figures (98 degrees).

A thermometer showing a reading of 98.3 degrees.

Some laboratory balances read to the hundredths place. The digital readout of an object's mass might look like this: 14.49. This reading has four significant figures. Other (more expensive) balances read to the ten-thousandths place. The digital display on this balance might read 14.4927 for this object. This reading has six significant figures.

1.4 What Are the Rules for Significant Figures?

How can you tell if digits are significant? Most of the confusion involves the number zero. Your textbook gives you rules similar to these.

All non-zero digits are significant.
    This means that any number that is not zero is significant. 7.121486 has seven significant figures; 4215 has four significant figures.

All zeros between non-zero digits are significant.
7002 has four significant figures; 7.1002 has five significant figures; 702 has three significant figures; 80.003 has five significant figures.

All zeros to the left of the first non-zero digit are not significant.
Here is where your normal, everyday definition of "significant" may clash with the chemistry definition. The number 0.0025 has only two significant figures, as the zeros to the left of the two are not significant. That does not mean they don't have to be there. If you took them away, you would get .25, a very different number! A common dosage of Vasotec™, a blood pressure-lowering drug, is 0.0025 grams. If you ignored the zeros, the dosage would become .25 grams, 100 times too high and probably fatal to the patient. Those zeros have the very important job of being placeholders; however, they are not significant in the sense of expressing the accuracy of the number.
If we convert 0.0025 kilograms to grams, it becomes 2.5 grams. Because we changed the units, we no longer needed the left-hand zeros as placeholders.

When a decimal point is present, zeros to the right of the last non-zero digit are significant.
Thus, 0.2300 has four significant figures, 2.4900 has five significant figures, and 2.0 has two significant figures.

If there is no decimal point written, you cannot tell if the zeros to the right are significant. This rule can confuse us if we are not careful.
For instance, we cannot say how many significant figures whole numbers such as 8000 and 400 have. They have one significant figure; beyond that we cannot say.

Now let's look at some examples that involve all the rules.

*   7.0010 has five significant figures. (All the zeros are significant because two are between non-zero digits and the last is in a number that has the decimal point written.)

- 0.0005 has one significant figure. (The zeros to the left are not significant.)

  - 0.000500 has three significant figures. (The zeros to the left of the 5 are placeholders but are not significant. The zeros to the right are significant because a decimal point is shown.)

- 1.000500 has seven significant figures. (All the zeros are significant because the three zeros are between non-zero digits and the last two are to the right in a number that has the decimal point written.)

- 7000 has one significant figure. (No decimal point is shown, so you cannot tell whether the zeros are significant.)

- 7000. has four significant figures. (This time the decimal point is shown.)

See Appendix A of your text for a discussion on significant figures in logarithms.

1.5 Scientific Notation to the Rescue

Let's look at the number 800. Suppose it represents 800 grams. If 800 grams is just a good guess at the weight of an object, then the number has only one significant figure. However, if you weighed the object and it weighed exactly 800 grams, the number would have three significant figures. If you want to show that the number has three significant figures, you would have to write it with a decimal point included; that is, 800. grams. But it is easy for a decimal point at the end of a whole number to "get lost." However, we can easily show the number of significant figures in numbers such as this by converting them to scientific notation. If you convert 800 to scientific notation, it becomes 8×10^2.

If you want to show that it has three significant figures, you can write it as 8.00×10^2.

When you write a number in scientific notation, the number of significant digits is expressed by the coefficient.

- 7000 grams becomes 7×10^3 with one significant figure
- 7000 grams becomes 7.0×10^3 with two significant figures
- 7000 grams becomes 7.00×10^3 with three significant figures
- 7000 grams becomes 7.000×10^3 with four significant figures

Remember that the exponent plays no part in determining the number of significant figures. It merely tells us where to put the decimal point.

Appendix 3 of your text further describes scientific notation and its use in arithmetic and multiplicative operations.

1.6   Rules for Rounding Off

You may remember these rules from a math class in high school. To round off a number to a certain place, you look at the number in the next place to the right of the number. If that number to the right is less than 5, that is, if it is 0, 1, 2, 3, or 4, you leave the number alone. This is called rounding down. If the number to its right is 5 or greater, that is, if it is 5, 6, 7, 8, or 9, you increase the value of the number by one. This is called rounding up.

Let's round some numbers.

- 7.134 rounded to the hundredths place becomes 7.13, since the number in the thousandths place is 4, a number less than 5.

- 0.21029 rounded to the ten-thousandths place becomes 0.2103, since the number to the right of that place is 9, a number that is 5 or greater.

1.7   Significant Figures in Multiplication and Division

When you multiply or divide measurements, the number with the least number of significant figures limits your answer. Suppose three students measure a box. Jane measures the length to be 3.12 cm, Carl measures the width to be 3.1 cm, and Donna measures the height to be 7.62 cm. To get the volume, we multiply the three sides together.

$$3.12 \text{ cm } [\times] \ 3.1 \text{ cm } [\times] \ 7.62 \text{ cm } [=]$$

The display reads 73.70064, with the answer in cm^3

However, we must round the calculated answer to 74 cm^3 because Carl made his measurement to only two significant figures.

When we calculate density, we divide mass by volume. That is,

$$\text{Density} = \frac{\text{Mass}}{\text{Volume}}$$

Let's say we measure a sample's mass to be 7.4219 grams and its volume to be 1.23 mL. We divide 7.4219 by 1.23.

$$\text{Density} = \frac{7.4219 \text{ g}}{1.23 \text{ mL}}$$

$$\text{Density} = 6.034065041 \ \text{g}\big/\text{mL}$$

Because 1.23 mL has only three significant figures, we must round the answer to 6.03 g/mL.

Significant figures are important, but don't get too bogged down with them. They will come naturally to you as you do more calculations.

1.8 [EXP]: Scientific Notation on the Calculator

Calculators make scientific notation easy. You just need to have the right calculator and know which buttons to push.

Note: Scientific calculators have a button that is marked EE (on most Texas Instruments calculators) or EXP (on most other brands). This is the button that lets you enter a number in scientific notation.

Let's take the number 7.45×10^8. To enter this number, we enter the coefficient 7.45, then press [EXP]. Two zeros appear at the right (smaller in size). These zeros are holding the places for the two digits of the exponent. (You will never need more than two.) We then enter the exponent 8. The display should look like this: 7. 45 ^{08}

When you see this display, you read it as 7.45×10^8. Press [C] to clear the display, and let's enter it again.

7.45 [EXP] 8   The display reads 7. 45 ^{08}

Note: A common mistake students make is to hit [X] before [EXP]. Don't do this! The "times 10 to the" is all included in [EXP].

So you *do not* enter

7.45 [×] [EXP] 8   or   7.4501 [×] 10 [EXP] 8

Simply press 7.45 [EXP] 8

Another example is to enter 9.12×10^{12}

Press 9.12 [EXP] 12   The display reads 9. 12 ^{12}

For an entry with negative exponent, try 7.216×10^{-9}

Enter 7.216 [EXP] 9 [±]   or   7.216 [EXP] [±] 9
The display reads 7. 216 $^{-09}$

The [±] key toggles a number between negative and positive. If you press [±] after [EE] or [EXP], it toggles the exponent between negative and positive. Press [±] a couple of times and see how the sign of the exponent changes back and forth.

Note: The [±] key is not the same as [–]. The [±] toggles between negative and positive. [–] is used only for subtraction.

Let's enter a few more numbers in scientific notation. Enter the mass of an electron, which is 1.6606×10^{-24} kilograms.

Press 1.6606 [EXP] 24 [±]

The display reads 1. 6606 ⁻²⁴

The second example is to enter 10^5. On your calculator:

Press 1 [EXP] 5   The display reads 1. 0 ⁰⁵

Note: When entering an exponential number where no coefficient is shown, you must supply the missing coefficient of 1. We can write 10^5 as 1×10^5.

Finally, let's enter 10^{-7}

Press 1 [EXP] 7 [±]   The display reads 1. 0 ⁻⁰⁷

1.9   Multiplication and Division with Scientific Notation

If you multiply two numbers in scientific notation, just press [×] between the numbers. Let's multiply $(2.12 \times 10^4)(7.89 \times 10^{-9}) =$

Enter 2.12 [EXP] 4 [×] 7.89 [EXP] 9 [±] [=]

The display reads 1. 67268 ⁻⁰⁴

Since each number has three significant figures, we round the answer to 1.67×10^{-4}. Note that the only time you enter [×] is between the two numbers.

Let's divide two numbers: $\dfrac{8.81 \times 10^4}{7.472 \times 10^8}$

Enter 8.81 [EXP] 4 [÷] 7.472 [EXP] 8 [=]
The display reads 1.1790685 $^{-04}$

Since the least number of significant figures is three, we round our answer to 1.18×10^{-4}.

Let's try another: $\dfrac{10^{-14}}{10^{-9}}$ (Remember to supply the missing coefficients of 1.)

Press 1 [EXP] 14 [±] [÷] 1 [EXP] 9 [±] [=]
The display reads 1.0000000 $^{-05}$, written 1×10^{-5} or 10^{-5}.

Now let's look at one last example that uses Avogadro's number (6.02×10^{23}), a number that you will see later in the class.

$$2.50 \times 10^{15} \left(\frac{12.0}{6.02 \times 10^{23}} \right)$$

Press 2.5 [EXP] 15 [×] 12 [÷] 6.02 [EXP] 23 [=]
The display reads 4.9833887 $^{-08}$.

Since each of the three numbers has three significant figures, we round the answer to 4.98×10^{-8}.

1.10 Switching between Scientific and Decimal

Sometimes you may want to convert an answer back and forth between scientific notation and decimal form. This will be most important to you on exams, especially multiple-choice exams where the answer choices could be written in either form.

The answers to problems in the back of your textbook may also be given in either scientific notation or decimal form. Often students are confused when they look up a hard-fought answer in the answer key and discover their answer looks wrong. Their answer may not be wrong. The textbook may have given the answer in scientific notation, while the student calculated it in decimal form.

Texas Instruments Calculators

On Texas Instrument calculators, we use [FLO] and [SCI] along with [2nd] to convert between scientific notation and decimal form. On the TI-30X calculator, [FLO] and [SCI] are second functions written in blue above [4] and [5]. We use [2nd] to access these operations.

Pushing [2nd] [FLO] puts the calculator into normal decimal mode. (FLO comes from "floating point," computer programmers' jargon for a decimal number. In this mode, any number we enter will change to decimal form as soon as we hit another key (as long as the number will fit the display).

Pushing [2nd] [SCI] puts the calculator into scientific notation mode. Any number we enter will change to scientific notation as soon as we hit another key.

First let's change some numbers in scientific notation to decimal form. Suppose you got 7.7×10^{-4} as an answer and want convert it to decimal form.

Enter 7.7 [EE] 4 [±]　　　The display reads 7.7 ⁻⁰⁴

Press [2nd] [FLO]　　　The display reads 0.00077

Now let's convert 8.449×10^{7} to decimal form.

Enter 8.449 [EE] 7   The display reads 8.449 ⁸⁷

Press [2nd] [FLO]   The display reads 84490000

Now let's try to convert 6.419×10^{-17} to decimal form.

Enter 6.419 [EE] 17 [±] The display reads 6.419 ⁻¹⁷

Press [2nd] [FLO] The display still reads 6.419 ⁻¹⁷
Nothing happens, because a number in scientific notation with an exponent of −17 has too many decimal places to fit into the 10 spaces on the calculator. A number in scientific notation must have an exponent of 9 or less, either plus or minus, to fit into the 10 decimal places on a calculator. That is, a number must be between nine billion and nine-billionth.

Now let's take some decimal numbers and convert to scientific notation. Suppose you got 0.000788 for an answer and want to change it to scientific notation.

Enter 0.000788   The display reads 0.000788

Press [2nd] [SCI]   The display reads 7.88 ⁻⁰⁴

Convert 6,992,000 to scientific notation.

Enter 6,992,000   The display reads 6992000

Press [2nd] [SCI]   The display reads 6.992 ⁰⁶

You can press [2nd] [SCI] to convert back to normal decimal form.

Note: If you press [2nd] [SCI] to convert an answer to scientific notation, the calculator remains in that mode until you press [2nd] [FLO] to put the calculator into normal decimal mode. If you enter a decimal number, it will change to scientific notation as soon as you press another function key such as [×], [÷] or [=] .

Likewise, if you are in decimal mode when you enter a number in scientific notation, it will change to decimal form as soon as you hit another key (as long as the decimal form will fit in the display). In either of these cases the number is not affected, only the form of the number.

Let's look at a few examples. Press [2nd] [SCI] to put the calculator in scientific notation mode

Enter 0.0059   The display reads 0.0059

Press [×] or [÷]

The display changes to 5.9000000^{-03}

Press [C] to clear the display. Now press [2nd] [FLO] to change your calculator to normal decimal mode.

Now enter 0.0059   The display reads 0.0059

Now press [×] or [÷]   The display still reads 0.0059

Casio, Sharp, and Most of the Rest

Most brands of calculators have a [MODE] button, usually in the upper left hand corner next to [INV]. They also have a table printed under the readout display that looks something like this.

MODE	•	0	4	5	6	7	8	9
	SD	COMP	DEG	RAD	GRA	FIX	SCI	NORM

or this

MODE	0:COMP	4:DEG	5:RAD	6:GRA
	• SD	7:FIX	8:SCI	9:NORM

The two numbers you need in this table to convert between scientific notation and decimals are 8:SCI (scientific notation) and 9:NORM (for normal decimal form).

Pressing [MODE] [9] puts the calculator into normal decimal mode. In this mode, a number you enter will change to decimal form as soon as you hit another key (as long as the number will fit in the display). Pressing [MODE] [8] [8] puts the calculator into scientific notation mode. Any number you enter will change to scientific notation as soon as you hit another key. The second number, the one you enter after pressing [MODE] [8], tells the calculator how many places of the coefficient the calculator will display. For instance, if you press [MODE] [8] [4], the answer will be rounded to four digits. It's better to let the calculator display eight digits; then you can choose the proper number of significant figures for an answer.

Let's change some numbers from scientific notation to normal decimal form. Suppose you got 7.7×10^{-4} as an answer and want to change it to decimal form.

Enter 7.7 [EXP] 4 [±]   The display reads 7. 7 $^{-04}$

Press [MODE] 9   The display reads 0. 00077

Convert 2.568×10^{8} to decimal form.

Enter 2.568 [EXP] 8   The display reads 2. 568 ^{08}

Press [MODE] 9   The display reads 256800000

Now let's try to convert 7.489×10^{-13} to decimal form.

Enter 7.489 [EXP] 13[±]
        The display reads 7. 489 $^{-13}$

Press [MODE] 9   The display still reads 7. 489 $^{-13}$

Nothing happens, because a scientific notation number with a −13 exponent has too many decimal places to fit into the 10 spaces on the calculator. A number in scientific notation must have an exponent of 9 or less, either plus or minus, to fit into the 10 decimal places on a calculator. That is, a number must be between nine billion and nine-billionth.

Now let's convert decimal numbers to scientific notation. Suppose you got 0.00489 as an answer and want to change it to scientific notation.

Enter 0.00489   The display reads **0. 00489**

Press [MODE] 8 8
          The display reads   **4. 8900000 ⁻⁰³**

Convert 2,090,000 to scientific notation.

Enter 2,090,000   The display reads **2090000**

Press [MODE] 8 8   The display reads **2. 090000 ⁰⁶**

Press [MODE] 8 8 to put the calculator in scientific notation mode.

Note: If you press [MODE] 8 8 to convert an answer to scientific notation, the calculator remains in that mode until you press MODE 9 to change to normal decimal mode. If you enter a decimal number, it will change to scientific notation as soon as you press another function key such as [x] or [÷].

Likewise, if you are in decimal mode when you enter a number in scientific notation, it will change to decimal form as soon as you press another key (as long as the decimal form will fit in the display). In either of these cases the number is not affected, only the form of the number.

Let's look at some examples. Press [MODE] 8 8; this puts the calculator in scientific notation mode;

Enter 0.0059   The display reads 0.0059

Press [×] or [÷]
       The display changes to 5.9000000 ⁻⁰³

Press [C] to clear the display. Now press [MODE] 9 to return your calculator to normal decimal mode.

Now enter 0.0059   The display reads 0.0059

Now press [×] or [÷]   The display still reads 0.0059

1.11   Does the Answer Make Sense?

A calculator is a very handy instrument. It will make computations quick and easy, but it is only as good as the numbers you enter. Computer programmers have an acronym for this: GIGO. Garbage in, garbage out. If you are in a hurry on a test, it is easy to push the wrong button and not notice it. Let's look at some calculations and see if we can tell if the answer makes sense.

$$\frac{9.2 \times 10^8}{5.5 \times 10^5} = 1.7 \times 10^3$$

We can check the answer to see if it's in the right ballpark by looking at the exponents: $10^8 + 10^5 = 10^3$. Also notice that 9.2 is a little less than twice 5.5, so the ratio of 1.7 is reasonable. Notice that the numerator is larger than the denominator, so the answer has a positive exponent.

Let's look at another example:

$$\frac{7.9 \times 10^{-4}}{6.9 \times 10^{-9}} = 1.1 \times 10^5$$

We look at the exponents: $10^{-4} \div 10^{-9} = 10^{-4-(-9)} = 10^{-4+9} = 10^5$. It checks out. In this case, we divided a larger number by a smaller number, and we get a number greater than one. Also notice that 7.9 is just a little larger than 6.9, so a ratio of 1.1 makes sense.

If we had hit the [×] button instead of the [÷] button by mistake, we would have gotten the wrong answer 5.5×10^{-12}. By quickly checking the exponents, we can spot this kind of error. The few seconds you spend checking your answers is time well spent.

Remember that studies show that most students can raise their grades one or even two letters by doing nothing more than being neat and checking their work.

Look at the following problem and the incorrect answer given.

$$\frac{7.50 \times 10^9}{\left(6.5 \times 10^7\right) \times \left(3.5 \times 10^3\right)} = 4.0 \times 10^7 \quad ???$$

Do a quick check of the exponents. Did your check tell you that the exponent in the answer should be around 10 and not 10^7? ($10^9 \div 10^5 \div 10^3 = 10^{(9-5-3)} = 10^1$)

Can you figure out what the student did wrong to arrive at the incorrect answer? The student carried out the calculation *incorrectly* as:

7.5 [EXP] 9 ÷ 6.5 [EXP] 5 x 3.5 [EXP] 3 [=]

instead of the correct key sequence:

7.5 [EXP] 9 ÷ 6.5 [EXP] 5 ÷ 3.5 [EXP] 3 =

Remember that on previous pages, we noted that all numbers in the denominator must be preceded by ÷ even if a × sign appears in front of the number. This student got the wrong answer because she forgot this rule.

On multiple-choice tests, the wrong answer you're likely to get is often one of the choices. Let's look at one last example.

$$12 \times \frac{1}{1,000} = 12,000 \quad ???$$

Here the student probably pressed [×] instead of [÷]. Since 1,000 is in the denominator, it will make 12 one thousand times smaller instead of one thousand times larger.

1.12 Practice Problems

Use your calculator. Round all answers to the appropriate number of significant figures. Answers are on the next page.

1. $\dfrac{4.807 \times 1.85}{0.467 \times 5.00} =$

2. $\dfrac{96.8}{15.2 \times 0.0120} =$

3. $10.7 \times \dfrac{1.2}{3.8} \times \dfrac{14.2}{5.01} =$

4. $\left(4.16 \times 10^5\right)\left(5.08 \times 10^{-8}\right) =$

5. $\dfrac{7.00 \times 10^{-6}}{5.871 \times 10^4} =$

6. $\dfrac{5.68 \times 10^3}{5.02 \times 10^2} =$

7. $\left(10^5\right)\left(10^{-8}\right)\left(10^3\right) =$

8. $\dfrac{10^{-14}}{10^{-9}} =$

9. $\left(3.15 \times 10^{18}\right) \times \left(\dfrac{342}{6.02 \times 10^{23}}\right) =$

10. $\left(1.00 \times 10^{15}\right) \times \left(\dfrac{1.66 \times 10^{-24}}{12.0}\right) =$

1.13   Answers to Practice Problems

1. 4.807 [×] 1.85 [÷] 0.467 [+] 5 [=]   3.808543897
    Rounded to 3 sig figs, the answer is 3.81

2. 96.8 [÷] 15.2 [÷] 0.012 [=]   530.7017544
    Rounded to 3 sig figs, the answer is 531

3. 10.7 [×] 1.2 [÷] 3.8 [×] 14.2 [÷] 5.01 [=]   9.577056413
    Rounded to 2 sig figs, the answer is 9.6

4. 4.16 [EXP] 5 [×] 5.08 [EXP] 8 [±] [=]   0.0211328
    Rounded to 3 sig figs, the answer is 0.0211
    In scientific notation, the answer is 2.11×10^{-2}

5. 7 [EXP] 6 [±] [÷] 5.871 [EXP] 4 [=]   1.1923011^{-10}
    Rounded to 3 sig figs, the answer is 1.19×10^{-10}

6. 5.68 [EXP] 3 [÷] 5.02 [EXP] 2 [=]   11.31474104
    Rounded to 3 sig figs, the answer is 11.3
    In scientific notation, the answer is 1.13×10^{1}

7. 1 [EXP] 5 [×] 1 [EXP] 8 [±] [×] 1 [EXP] 3 [=]   1
    The answer is simply 1

8. 1 [EXP] 14 [±] [÷] 1 [EXP] 9 [±] [=]   0.00001
    The answer in scientific notation is 1×10^{-5}

9. 3.15 [EXP] 18 [×] 342 [÷] 6.02 [EXP] 23 [=]   1.7895349^{-03}
    Rounded to 3 sig figs, the answer is 1.79×10^{-3}
    The decimal form is 0.00179

10. 1 [EXP] 15 [×] 1.66 [EXP] 24 [±] [÷] 12 [=]   1.38333^{-10}
    Rounded to 3 sig figs, the answer is 1.38×10^{-10}

SECTION 2

Skills for Chapter 2: Measurement and Problem Solving

2.1 Scientific Notation

Scientific notation is a method used to express very large and very small numbers on your calculator. These numbers are expressed by multiplying the significant portion of the number by a multiplier based on 10^x.

To use the scientific notation feature on your calculator, you will use a [sci], [exp], or [EE] function. Consider the process of entering 1.6×10^{-17}. First enter the significant portion of 1.6 into the display. Next press the [sci] or equivalent key and enter the number 17. You will need to change the sign of the power of 17 (which should appear in the right-hand portion of the display), so press the [+/−] key to cause the power to read −17. Depending on the type of calculator you have, you may need to press the [=] key to end the sequence.

You will also be able to convert from regular notation to scientific notation on your calculator. If the number 1,000,000 was in the display of your calculator, you could depress the [sci] then [=] keys to cause the number to appear in scientific notation (1.000×10^6) on the display.

An example of scientific notation that is found in Chapter 9 of your textbook involves the determination of the frequency of a photon from its wavelength. For instance, if a HeNe laser has a wavelength of 632.8 nm, what is the frequency of the photon? The answer is calculated using $c = \nu\lambda$, where c is the speed of light (a constant), ν is the frequency (in Hertz), and λ is the wavelength in meters. We can quickly estimate the frequency by expressing all numbers in scientific notation and setting up the problem. Using this, we write:

$$c = 2.998 \times 10^8 \text{ m/s}$$
$$\lambda = 632.8 \text{ nm or } 632.8 \times 10^{-9} \text{ m}$$
$$\text{or} \quad 6.328 \times 10^{-7} \text{ m}$$

The problem can be set up to solve for v as:

$$v = \frac{c}{\lambda}$$

$$v = \frac{2.998 \times 10^8 \text{ m/s}}{6.328 \times 10^{-7} \text{ m}}$$

$$v = 4.738 \times 10^{14} \text{ Hz}$$

Before using the calculator, we can get a rough idea of the magnitude of the number. The exponential portion of the calculation shows a +8 in the numerator and a −7 in the denominator. Using our algebra skills, we can determine the magnitude of the exponent is +15. The non-exponential term of the equation can be reduced to 3/6, which is 1/2 (or 0.5). Hence, our rough guess of the answer is 0.5×10^{15} Hz, or 5×10^{14} Hz.

The calculator would allow us to find the answer of 4.738×10^{14} Hz, based on four significant digits. This answer is close to our estimate. More examples of scientific notation can be found in Examples 2.1 and 2.2, Skillbuilders 2.1 and 2.2, and Problems 27–36.

2.2   Significant Figures

Our calculators normally display six to nine digits when performing arithmetic functions. While it may seem advantageous to have such large numbers when performing chemical computations, not all of this information may be significant because of error associated with the data.

To put this into context, consider the data that you record in the laboratory. Every measurement made is subject to error. The level of this error (or certainty) depends on the instruments used in the measurement and the skill of the person performing the operations. Even if we can eliminate the systematic errors (i.e., miscalibration of the pan balance), we will still encounter random errors in lab. These random errors will determine the accuracy of our measurement.

Rather than stating the error with each number a chemist may use in a calculation, the practice of using "significant digits" in calculations is employed. The significant digits of a number can be thought of as those digits in the number that do not change when the uncertainty is factored in. (The rules used for determining the number of significant digits in an expressed value are listed in Section 2.3 of your textbook.) As an example of the effect of uncertainty on the number of significant digits, consider a food sample, which was found to have 1.33 mg of Ca per serving, with an error of 0.1 mg for the determination. Based on the uncertainty, we should only say that the sample contains 1.3 mg of Ca. The number has only two significant digits. It is of no importance to attempt to convey that the calculation indicated a second 3 in the decimal. The error made this digit "insignificant." For practice in significant digits and rounding off nonsignificant digits, see Examples 2.3, 2.4, and 2.21, Skillbuilders 2.3 and 2.4, and Problems 37–44. Problems 45–50 deal with the rounding of numbers.

You will be asked in this course to use significant digits in the estimation of answers to your calculations. As an example, consider the problem of determining the circumference of a cylinder. With a simple ruler you could determine the diameter to be 2.5 inches. From the relationship of $C = \pi \times d$, you could use your calculator to determine the circumference, C. The display of the calculator would read (7.853981...). What value do you report? What is significant?

The basic rule is that you report your answer using the least number of significant digits in the input data. In the multiplicative operation above, the diameter was only reported to two significant digits. The answer, therefore, should be reported as 7.9 inches. See Examples 2.5 and 2.22, Skillbuilder 2.5, and Problems 51–54 for more work with multiplicative operations and significant digits.

Rules for addition and subtraction are slightly different. Consider adding a 0.001 g weight to an object resting on a pan balance and indicating a value of 23.2 g in mass. The addition of 0.001 to 23.2 would yield a theoretical answer of 23.201 g. However, if the pan balance used had an error of 0.1 g, the addition of 0.001 g to this weight would be insignificant. The uncertainty will limit our ability to express accurately the sum of the combined weights. See Examples 2.6 and 2.23, Skillbuilder 2.6, and Problems 55–58 for work with significant figures for addition and subtraction.

Example 2.7 and Skillbuilder 2.7 show the correct use of significant figures in calculations that use multiplication, division, addition, and subtraction. Example 2.24 and Problems 57–62 also address this topic.

2.3 Mean and Standard Deviation

The uncertainty associated with data is often not evident from just looking at the data. Instead, the uncertainty is calculated based upon all the data you have available. The basic calculations are the mean and standard deviation. These two values can be used to determine the accuracy and precision of your measurements.

If your calculator has statistical analysis functions, you can calculate the mean ($\bar{x}$) and standard deviation (s) of a data set by simply entering the data and pressing the appropriate function key. As an example, let us say you titrated five equal volumes of an unknown acid with a standard base. You have dispensed the following volume of titrant in the five runs: 43.80 mL, 43.30 mL, 44.10 mL, 43.90 mL, and 43.70 mL. We would first enter the five values into the memory of the calculator. After the final entry, we would depress the button to give us the mean, $\bar{x}$. The value of 43.76 should appear. Next we would calculate the standard deviation, s, by pressing the appropriate buttons (s, or s_x). A value of 0.30 should appear.

Based on your titrating skills, you could say you dispensed 43.76 ± 0.30 mL per titration in this data set. Considering the error in the data (0.30 mL), expressing a value beyond the tenths of mL (0.1 mL) would not be significant. The correct answer should be 43.8 ± 0.3 mL, where only three significant digits are reported.

SECTION 3

Skills for Chapter 3: Matter and Energy

Goals: To perform calculations with specific heat.
Skills: Multiplication and division.

The concept of specific heat is outlined in Chapter 3.10 of your text. This term allows you to calculate the heat required to raise the temperature of one gram of a specific element, compound, or material. Conversely, this term will allow you to determine the heat that is released or absorbed by a chemical reaction if the temperature change of the reaction can be measured.

As an example of this concept, consider the heat q that is generated when solutions of acids and bases are mixed together. The heat evolved in a reaction can be expressed using the following equation:

$$\text{Heat} = \text{Mass} \times \text{Heat capacity} \times \text{Temperature change}$$

$$q = m \times C \times \Delta T$$

The mass for the mixture (m) is determined by weighing or measuring the volume of liquid in the calorimeter and multiplying this volume by the density of the solution. The heat capacity C is normally listed in a table. The ΔT is calculated from the temperature readings. The ΔT term is defined as the final temperature minus the initial temperature. A reaction where heat evolved (exothermic) will have a positive ΔT while a reaction where heat is absorbed from the environment (endothermic) will have a negative ΔT.

For example, when the acid and base were mixed in the calorimeter, the temperature of the solution changed from 23.5°C to 25.9°C. The final volume of the mixture was 50.0 mL. Since the density of the solution (being mostly water) is 1.00 g/mL, the aqueous solution weighs 50.0 g. The specific heat of water is 4.18 J/°C·g. The

ΔT is calculated as $25.9°C - 23.5°C = 2.4°C$. Placing these values into the above equation yields:

$$q = (50.0 \text{ g}) \times (4.18 \text{ J/°C·g}) \times (2.4°C)$$
$$q = 500 \text{ J}$$

See Examples 3.9, 3.10, and 3.18 and Skillbuilders 3.9 and 3.10 for problems using heat capacity. Problems 65–80 involve work with heat capacity, as well as energy and temperature change.

A point to remember is that any expression that involves reactions will deal with the difference in the final state from the initial state. For any calculation using Δ, (i.e., ΔT), make a habit of thinking that $\Delta = $ (final state – initial state). Using this approach you won't get the sign of the Δ wrong.

SECTION 4

Goals: To calculate yields.
Skills: Multiplication and division.

P art of your studies dealing with chemicals and their reactivity will be to determine the yield of a reaction. From your readings in Chapter 8 concerning stoichiometry, if we write the equation,

$$A + B \rightarrow C$$

we understand if 1 mol of A and 1 mol of B are allowed to react under favorable conditions, 1 mol of C should be produced. The *theoretical yield* of the reaction is 1 mol of C. This value could be expressed in grams instead of moles by multiplying the theoretical yield of 1 mol by the molar mass of the compound C.

Note that yield depends on the stoichiometry of the reaction. If the reaction were instead,

$$2A \rightarrow B$$

and 1 mole of A was placed in the reaction vessel, the theoretical yield would be 0.5 moles of B. Using numerical problem-solving number strategies (Chapter 2.10 of your text), the problem could be solved.

$$x \text{ mol} = 1 \text{ mol A} \times \frac{1 \text{mol B}}{2 \text{mol A}}$$

$$x \text{ mol} = 0.5 \text{ mol B}$$

Example 8.4 involves the calculation of the theoretical yield from the limiting reagent in a reaction. Also, see Skillbuilder 8.4 and Problems 41–44, 47, 48, 53, and 54. The determination of the yield based on the limiting reagent is the subject of Examples 8.5 and 8.6, Skillbuilders 8.5 and 8.6, and Problems 45, 46, 51, and 52.

In practice, the mol quantity of product that is obtained from the reaction is less than the theoretical value. The amount of product that is obtained is termed the *actual yield.*

A method in which the efficiency of a reaction is gauged is the calculation of the *percent yield.* It is defined as

$$percent\ yield = \frac{actual\ yield}{theoretical\ yield} \times 100\%$$

For the first reaction, if we had obtained 0.8 mol of C, the % yield would be

$$percent\ yield = \frac{0.8\ mol}{1\ mol} \times 100\%$$

$$percent\ yield = 80\%$$

The calculation of percent yield can also be performed using the grams of actual product obtained as compared to the theoretical gram yield. Problems with percent yield are 55 and 56.

In many cases, the scientist may wish to predict the actual yield for a well-characterized reaction. For instance, if the percent yield of a synthetic reaction is 90%, the actual yield may be found by multiplying the percent yield by the theoretical yield. (Remember to divide the percent yield by 100% in order to remove the % unit from the expression.) Example 8.6 and Skillbuilder use this approach to determine the actual yield of a reaction. Example 89 shows the calculation of limiting reagents, theoretical yields, and percent yields. Additional problems are 57–62.

SECTION 5

Skills for Chapter 11: Gases

Goals: To perform calculations with gas laws and the ideal gas law.
Skills: Cross multiplication and division.

5.1   Gas Laws

In this section of your studies, you will be exploring relationships between pressure, P, volume, V, temperature, T, and mol quantities of gases, n. Your text will introduce you to Boyle's law, Charles's law, Gay-Lussac's law, and the combined gas law.

You must exercise care in two areas to avoid mistakes when using these laws. The first area is temperature: all temperatures must be expressed in Kelvin (K) even if the data is given to you in °C. Remember, the relationship is K = °C + 273.15. The second area is to rearrange the equation so that the left-hand side only contains the term that you are seeking. All the given data will then appear on the right-hand side of the equation.

Consider the use of Charles's law in finding the volume that a gas in a balloon would occupy if it was heated from 25°C to 75°C. (This law is discussed in Chapter 11.5 of your textbook.) The initial volume of the gas (V_1) at 25°C is 1.00 L. First, let's set up the equation:

$$\frac{V_1}{T_1} = \frac{V_2}{T_2}$$

From the above information we can write:

$$V_1 = 1.00 \text{ L}$$
$$T_1 = 25°C, \text{ which is } 298 \text{ K}$$
$$V_2 = ?$$

$$T_2 = 75°C, \text{ which is } 348 \text{ K}$$

To solve for V_2, we write:

$$\frac{V_1T_2}{T_1} = V_2 \qquad \text{(cross multiplying } T_2)$$

$$V_2 = \frac{V_1T_2}{T_1} \qquad \text{(switching sides)}$$

By plugging in the data, we obtain:

$$V_2 = \frac{1.00 \text{ L} \times 348 \text{ K}}{298 \text{ K}}$$

$$V_2 = 1.17 \text{ L}$$

When dealing with Boyle's, Charles's, or the combined gas law, you will often find a great deal of information in the problem. Follow these steps to solve such problems:

1) Write the gas law equation to be used.
2) Sort the given data (as set 1 or set 2).
3) Rearrange the equation to solve for the unknown term.
4) Enter the data and solve.

Problems with Boyle's Law are found in Examples 11.2, Skillbuilder 11.2, and Problems 37, 38, 41, and 42. Charles's law problems are found in Example 11.3, Skillbuilder 11.3, and Problems 39, 40, and 43–54. Problems with Avogadro's Law are shown in Examples 11.5 and 11.14, Skillbuilder 11.5, and Problems 53 and 54, while the combined gas law is covered in Examples 11.4 and 11.15, Skillbuilder 11.4, and Problems 55–62.

5.2 Ideal Gas Law

Another law that you will use is the ideal gas law, $PV = nRT$. It is described in Chapter 11.8 of your textbook. Here, n is the number of moles of the gas and R is the gas constant [(0.08206 L·atm)/(K·mol)]. This gas law will allow you to work any of the previous problems that relate the P, V, and T of a gas to the mole quantity of the gas. For instance, what volume would 1.00 g of CO_2 occupy at 1.00 atm of pressure and a temperature of 23°C? Begin this problem by writing the equation:

$$PV = nRT$$

We were not given n in the original problem, but we can calculate it by using the equation $n = g/molar\ mass$. Here,

$$n = \frac{1.00\ g}{44.0\ g/mol\ CO_2}$$

and is equal to 0.0227 mol. With this information we can write:

$$P = 1.00\ \text{atm}$$
$$V = ?$$
$$n = 0.0227\ \text{mol}$$
$$R = 0.08206\ (\text{L·atm})/(\text{K·mol})$$
$$T = 23°C, \text{ which must be expressed as 296 K}$$

Rearranging the equation to solve for V, we write:

$$V = \frac{nRT}{P}$$

Plugging in the data:

$$V = \frac{0.0227\ \text{mol} \times 0.08206\ (\text{L}\cdot\text{atm})/(\text{K}\cdot\text{mol}) \times 296\ \text{K}}{1.00\ \text{atm}}$$

$$V = 0.551 \text{ L}$$

Examples 11.6, 11.7, and 11.16, Skillbuilders 11.6 and 11.7, and Problems 63–72 make use of the ideal gas law for finding P, V, and T of gases.

You are also introduced to a set of conditions on page 376 of your text that define the standard temperature and pressure (STP) of a gas. These conditions are often used in determining the molar volume of a gas. The molar volume is found by using a pressure of 1.00 atm and a temperature of 273 K in the ideal gas law. As shown on this page of your text, one mole of gas occupies a volume of 22.4 L, regardless of the molecular weight of the gas. See Problems 80 and 90 for the calculation of molar volumes, and Problems 91 and 92 for the determination of the mass of a gas at STP.

Another important use of the ideal gas law involves the determination of the molar mass of a gas (see Chapter 11.8). The molar mass of a gas is the weight of one mole of gas. It can be determined by replacing the n term in the ideal gas law with m/MM, where m represents the grams of the gas and MM represent the molar mass of the gas. Example 11.8, Skillbuilder 11.8, and Problems 73–76 show the use of the ideal gas law to find the MM of an unknown gas.

5.3   Mixtures of Gases

The pressure of a gaseous mixture is dependent on the partial pressure exerted by each gas. Using kinetic molecular theory, the partial pressure of each gas is expressed as $P_A = X_A P_{tot}$, where P_A is the pressure of gas A, X_A is the mole fraction of gas A, and P_{tot} is the total pressure of the gas. To illustrate this law, consider a container where 1.0 mol of gas A is mixed with 3.0 mols of gas B. The mole fraction of gas A is calculated as follows:

$$X_A = \frac{1.0 \text{ mole of gas A}}{4.0 \text{ moles of gas total}}$$

$$X_A = 0.25$$

Similarly,

$$X_B = \frac{3.0 \text{ moles of gas B}}{4.0 \text{ moles of gas total}}$$

$$X_B = 0.75$$

IMPORTANT: The sums of all the mole fractions should always add up to 1. Since the mole fraction is a simple ratio, it has no units.

In working the problem, we are told the pressure in the container is 2.00 atm. What then is the partial pressure of gas A? The equation is:

$$P_A = X_A P_{tot}$$

The terms to use in the equation are:

$$X_A = 0.25 \qquad \text{(calculated above)}$$
$$P_{tot} = 2.00 \text{ atm} \qquad \text{(given in the problem)}$$

Now by substituting into the equation we obtain,

$$P_A = 0.25 \times 2.00 \text{ atm}$$
$$P_A = 0.50 \text{ atm}$$

By similar reasoning we could determine that $P_B = 1.50$ atm. Also, since the gas mixture consisted of only two gases, if the pressure of gas A is 0.50 atm and the total pressure of *both* gases is 2.00 atm, by the difference we can find the pressure of gas B to be 1.50 atm. Examples 11.9, 11.10, and 11.17, Skillbuilders 11.9 and 11.10, and Problems 77–80 contain similar work.

Note that Problems 81 and 82 deal with the determination of the pressure of a gas that is produced in chemical reaction. The gas is collected by bubbling it into a tube filled with water. The determination of the actual pressure of the produced gas requires us to subtract the P_{water} from the overall pressure reading. Other exercises with the pressures are found in Problems 83–86.

SECTION 6

Skills for Chapter 12: Liquids, Solids, and Intermolecular Forces

Goals: To calculate heat required for vaporization and fusion
Skills: Addition, multiplication, and division

6.1 Heat of Vaporization

The determination of heat required in vaporization processes involves the use of the standard enthalpies of formation. These values are found in Table 12.2 of your text. The (ΔH_{vap}) is the amount of heat required to vaporize one mole of a compound. For instance, the (ΔH_{vap}) of water is 40.7 kJ.

Calculations involving the heat of vaporization require that we know the mole quantity of the compound that undergoes vaporization. The general formula that we use is:

$$\text{heat} = (\Delta H_{vap}) \times n$$

where n is the number of moles of the compound that undergo vaporization.

Consider the amount of heat needed to vaporize 100 g of H_2O. Using the equation above, we write:

$$\text{heat} = \Delta H_{vap} \times \left(\frac{\text{mass}}{\text{molecular weight}} \right)$$

$$\text{heat} = 40.7 \text{ kJ/mol} \times \frac{100 \text{ g}}{18.01 \text{ g/mol}}$$

$$\text{heat} = 256 \text{ kJ}$$

Variations of this example may involve the determination of the mass of the compound if the heat is known. See Examples 12.1 and 12.7, Skillbuilder 12.1, and Problems 49–56 for work with heat of vaporization.

6.2 Heat of Fusion

The heat of fusion ΔH_{fus} is the amount of heat required to melt 1 mole of a solid. Values of ΔH_{fus} are shown in Table 12.3. The heat involved in this process is calculated in a very similar fashion to that of ΔH_{vap}. Calculations use the following formula:

$$\text{heat} = \Delta H_{fus} \times \left(\frac{\text{mass}}{\text{molecular weight}} \right)$$

See Examples 12.2 and 12.8, Skillbuilder 12.2, and Problems 57–60 for ΔH_{fus} for work with the heat of fusion.

SECTION 7

Skills for Chapter 13: Solutions

Goals: To calculate concentrations, dilutions, freezing point depression and boiling point elevation
Skills: Multiplication and division.

7.1 Concentration Units

A very important factor of solutions involves concentration units. In Chapter 13 of your text, you are introduced to two terms: mass percent and molarity. Both of these terms involve the measurements of the mass of the solute and the amount of solution.

The mass percent is a simple ratio of the mass of solute to the mass of the solution. The term "percent" involves the multiplication of the simple ratio by 100 %. Both masses are obtained by weighing the solute and the solvent. For instance, if we place 10 g of NaCl in a beaker with 100 g of water, we can say the mass of the solute is 10 g and the mass of the solvent is 100 g.

The calculation of the mass percent involves placing the mass of the solute in the numerator of the equation and the mass of the solution in the denominator. Since our solution contains 100 g of water and 10 g of NaCl, the mass of the entire solution is 110 g. This is shown below.

$$\text{mass percent} = \left(\frac{10 \text{ g of NaCl}}{100 \text{ g of water} + 10 \text{ g of NaCl}} \right) \times 100\%$$

$$\text{mass percent} = \left(\frac{10 \text{ g of solute}}{110 \text{ g of solution}} \right) \times 100\%$$

$$\text{mass percent} = 9.1\%$$

Problems that involve calculation with mass percent are Examples 13.1, 13.2, 13.11 and 13.12, Skillbuilders 13.1 and 13.2, and Problems 43-60. Note that Problems 51 and 52 involve the calculation of the mass of solute when you are given the mass of the solution. Problems 55 and 56 involve the density of the solutions. The work in Problems 59 and 60 deal with mass percent when the solute is expressed in units of mg.

The molarity, M, is the term that expresses the mol quantity of a dissolved substance (solute) per liter of solution. A solution that is 1.00 M NaCl contains 1.00 mol of NaCl per 1.00 L of solution. This same solution contains 58.4 g of NaCl per 1.00 L of solution.

Calculations that involve molarity will usually begin with the statement of the gram quantities of solute dissolved in a quantity (mL or L) of solution. By definition,

$$M = \frac{\text{mol of solute}}{\text{L of solution}} = \frac{\left(\frac{\text{g of solute}}{\text{MW of solute}}\right)}{\text{L of solution}}$$

For instance, to calculate the molarity of a solution that is 10.0 g of NaCl in 400 mL of water, we would write:

$$M = \frac{\left(\frac{10.0 \text{ g}}{58.4 \text{ g/mol}}\right)}{0.400 \text{ L}}$$

We will first solve the numerator of the problem for the amount of NaCl present in the solution. If we divide 10.0 g by 58.4 g/mol, we should obtain a value of 0.171 mol NaCl. Substituting into the equation

$$M = \frac{0.171 \text{ mol}}{0.400 \text{ L}}$$

$$M = 0.428 \text{ mol NaCl / L of solution}$$

Note: Express the volume of solution only in L (liters).

See Examples 13.3 and 13.3, Skillbuilder 13.3, and Problems 61-64 of your textbook for similar problems.

You may find it necessary in your work to determine what amount of solute, expressed in grams, would result in a certain molarity of solution. Consider what gram amount of NaCl is necessary to create a 0.60 M solution that is 750 mL in volume. First, let's fill the given information into the molarity expression.

$$0.60 \text{ M} \ = \ \frac{\left(\dfrac{x \text{ g}}{58.4 \text{ g/mol}}\right)}{0.750 \text{ L}}$$

To solve for x, we must cross-multiply several factors in order to arrange the equation for x.

$$(0.60 \text{ mol/L}) \times (0.750 \text{ L}) = \frac{x \text{ g}}{58.4 \text{ g/mol}}$$

$$(0.60 \text{ mol/L}) \times (0.750 \text{ L}) \times (58.4 \text{ g/mol}) = x \text{ g}$$

$$x \text{ g} = 26.3 \text{ g of NaCl}$$

Note that all units on the left side of the equation reduced to g (remember that M is in mol/L). By taking time with your cross multiplication step, the problem can be solved. This example is somewhat similar to Example 13.14 and Problems 67, 68 and 73-76.

Another use of the molarity equation involves the determination of the volume of solution needed to deliver a specific quantity of a chemical reagent. See Example 13.4, Skillbuilder 13.4, and Problems 69-72, 77 and 78 for work with this concept.

7.2 Dilution

Dilution involves reducing the concentration of a solution by adding more solvent. Typically, a volume of solution is taken and added to a volume of solvent. Consider the dilution of a 3.00 M NaCl stock solution by pipetting 10.0 mL into a 1.000 L flask and filling the flask to the calibration mark with pure water. Here we are adding roughly 990 mL of water to the solution. The concentration of the diluted solution is calculated as:

$$M_1 \times V_1 = M_2 \times V_2$$

$$M_2 = M_1 \times \frac{V_1}{V_2}$$

where M_2 is the final concentration of the diluted solution, M_1 is the initial concentration of the concentrated solution, V_2 is the final concentration of the diluted solution, and V_1 is the initial concentration of the solution. For our problem, $M_1 = 3.00$ M NaCl, $V_1 = 10.0$ mL, and $V_2 = 1.000$ L (or 1,000 mL). Substituting these numbers into the equation, we write:

$$M_2 = 3.00 \text{ M} \times \frac{10.0 \text{ mL}}{1000 \text{ mL}}$$

$$M_2 = 0.0300 \text{ M}$$

Note that the volume units in a dilution calculation must be the same no matter how they are stated in the problem. Use the numerical problem solving strategy (Chapter 2.10) to check your units before reporting your answers. See Problems 79 and 80 and Example 13.15 of your text for more dilution problems.

The dilution equation can also be used to solve for V_1, if the other variables of the equation are known. For instance, let's consider what volume (V_1) of a 16.0 M solution of HNO_3 (M_1) must be added to a beaker and diluted to 2.25 L (V_2) in order to produce a final solution of 0.10 M HNO_3 (M_2). Inserting these data into the dilution equation, we write:

$$V_1 = 2.25 \text{ L} \times \frac{0.10 \text{ M}}{16.0 \text{ M}}$$

$$V_1 = 0.141 \text{ L} \quad \text{or} \quad 14.1 \text{ mL}$$

Example 13.6 and Skillbuilder 13.6 is somewhat similar to this problem. Also see Problems 81-86 in your textbook for more practice in calculating dilution volumes.

7.3 Freezing Point Depression and Boiling Point Elevation

In Chapter 13.9, you are introduced to two chemical phenomena that are concentration dependent: the depression of the freezing point of a solution, and the elevation of the boiling point of a solution. These effects are determined by the molality of a nonvolatile solute in a solvent.

The molality concentration unit (m) is calculated as the ratio of the moles of solute per kg of solvent. Examples 13.7 and 13.8, Skillbuilders 13.7 and 13.8, and Problem 95-98 involve molality calculations.

The effect of a nonvolatile solute on the freezing point and boiling point uses a set of equations that appear to be similar.

$$\Delta T_f = m \times K_f \qquad \text{freezing point depression}$$

$$\Delta T_b = m \times K_b \qquad \text{boiling point elevation}$$

Note that each equation finds the ΔT for each effect. The final boiling point or freezing point is determined from the effect of ΔT.

The values of K_f and K_b are specific for each solvent. Only K_f values can be used for the freezing point depression and K_b for boiling point elevation. Example 13.9, Skillbuilder 13.9, and Problems 99 and 100 involve freezing point depression. Boiling point elevation is the subject of Example 13.10, Skillbuilder 13.10, and Problems 101 and 102. Problems 101-104 and Example 13.19 involve both K_f and K_b calculations.

SECTION **8**

Skills for Chapter 14: Acids and Bases

Goals: Calculations of acid-base titrations and pH.
Skills: Multiplication, division, and logarithms.

8.1 Titrations

As discussed in Chapters 14.6 of your textbook, this process involves reacting a standard solution with an unknown solution for the purpose of determining the concentration of the unknown solution. In the lab, commonly used titrations involve acid-base reactions, precipitation reactions, redox reactions, and complexation reactions.

Before any titration can be carried out, the solution stoichiometry must be known. Specifically, we must know how the titrant (standard solution) will react with the unknown. Using a simple acid-base reaction that is listed below, the mathematics for a titration can be set up.

$$HCl(aq) + NaOH(aq) \rightarrow H_2O(l) + NaCl(aq)$$

In this reaction, one mole of HCl will reacts with one mole of NaOH. This relationship allows the following equation to be written:

$$mol_{acid} = mol_{base}$$

Substituting the expression of equivalent $= M \times V$, we obtain:

$$M_{acid} \times V_{acid} = M_{base} \times V_{base}$$

Consider a titration of an unknown NaOH solution with a standardized HCl solution. To 50.0 mL of the unknown contained in a flask, approximately 33.8 mL of a 0.175 M standardized HCl solution was added to the flask to reach the equivalence point. From this information, we can determine the normality of the unknown solution.

If we rearrange the above equation by cross multiplication and division, we can solve for the normality of the unknown base, N_{base}. Doing so yields:

$$M_{base} = \frac{M_{acid} \times V_{acid}}{V_{base}}$$

Inserting the data (with volumes in L), we find:

$$M_{base} = \frac{0.175 \text{ mol/L} \times 0.0338 \text{ L}}{0.0500 \text{ L}}$$

$$M_{base} = 0.118 \text{ M}$$

Compare this problem to Examples 14.4 and 14.13 and Problems 51 and 52 of your text.

For an example where the solution stoichiometry is different, see Skillbuilder 14.4 and Example 14.13 of your text. In the Skillbuilder H_2SO_4 is used. This diprotic acid reacts with 2 moles of the base KOH. The solution to this type of titration problem requires starting out with the equation:

$$1 \text{ mol}_{acid} = 2 \text{ mol}_{base}$$

$$M_{acid} \times V_{acid} = 2(M_{base} \times V_{base})$$

As with the previous problem, the M_{acid} can be determined from the volumes of acid and base used in the titration and the normality of the KOH solution. Compare this example to Problems 53 and 54 of your text.

8.2   pH Calculations

Your text discusses the concept of pH in Chapter 14.9. The unit pH is a most important term when dealing with acid-base chemistry because it describes to us the amount of H^+ ions that are present (and potentially reactive) in an aqueous solution.

When HCl is added to H_2O, the following reaction occurs:

$$HCl(aq) + H_2O(l) \leftrightarrow H_3O^+(aq) + Cl^-(aq)$$

Instead of simply writing H^+ as one of the products of this reaction, we instead write $[H_3O^+]$. As your text mentions, the production and concentration of $[H_3O^+]$ is controlled by the dissociation of water. Water is capable of producing $[H_3O^+]$ according to:

$$2 H_2O(aq) \leftrightarrow H_3O^+(aq) + OH^-(aq) \qquad K_w = 1.0 \times 10^{-14}$$

A solution consisting only of water will have a H_3O^+ concentration of 1×10^{-7} M. If a strong acid such as HCl is added to the water, the $[H_3O^+]$ will increase.

Instead of always trying to remember the hydrogen ion concentration of an acid solution as a value expressed to a power, chemists have simplified the values through the concept of pH. By definition:

$$pH = -\log[H_3O^+]$$

If you know the H_3O^+ equilibrium concentration, the pH may be obtained by entering the value of $[H_3O^+]$ into your calculator, pressing the [*log*] key, and then the [+/−] to make the value positive. For example, if the $[H_3O^+]$ is found to be 3.0×10^{-5} M, then the pH is calculated as:

$$pH = -\log[3.0 \times 10^{-5}]$$
$$pH = -(-4.52)$$
$$pH = 4.52$$

Important: There are two types of logarithmic functions on your calculator, [*log*] and [*ln*]. The pH definition is based on *log* and not *ln*.

We can check the pH calculations with a given $[H_3O^+]$ by using the *log* relationship of powers to pH.

$$[H_3O^+] = 1.0 \times 10^{-3}$$

$$pH = -\log[1.0 \times 10^{-3}]$$

$$pH = -(-3.00)$$

$$pH = 3.00$$

By similar reasoning, if $[H_3O^+] = 1.0 \times 10^{-5}$, then the pH = 5.00. For the first example worked in this section where $[H_3O^+] = 3.0 \times 10^{-5}$ M, we can predict that the pH should lie between 4 and 5. This second check with our "cerebral calculator" can be of aid in confirming the answer from our calculator. Similar problems are shown in Examples 14.8 and 14.14, Skillbuilder 14.8, and in Problems 73–74 of your text.

Another calculation that will be used in these studies is the determination of the $[H_3O^+]$ if we are given the pH. This calculation can be performed by using the concept of *antilogs*. From our earlier definition,

$$pH = -\log[H_3O^+]$$

$$-pH = \log[H_3O^+] \qquad \text{(multiplying both sides by -1)}$$

$$10^{-pH} = 10^{\log[H_3O^+]} \qquad \text{(using the \textit{antilog} relationship)}$$

$$10^{-pH} = [H_3O^+]$$

$$[H_3O^+] = 10^{-pH}$$

To perform these calculations, use the $[y^x]$ or the $[10^x]$ function on your calculator. If the pH is 5.5, then:

$$[H_3O^+] = 10^{-5.5}$$

$$[H_3O^+] = 3.2 \times 10^{-6} \text{ M}$$

Similar calculations are shown in Examples 14.9, 14.18, and Skillbuilder 14.9 of your textbook. Problems 75 and 76 reinforce this concept.

Note that several problems involve the calculation of the pH if you are given the OH^- concentration. The relationship between H_3O^+ and OH^- is discussed in Chapter 14.8 of your textbook. From this relationship, we know that:

$$pK_w = [H^+] \times [OH^-]$$

and with the K_w of water at 25°C being 1×10^{-14}

$$14 = [H^+] \times [OH^-]$$

By knowing the $[OH^-]$, you can calculate the $[H^+]$ and then the pH. See Example 14.16.

Of course, you can also use the mathematics introduced in this chapter to calculate the pOH and relate it to the pK_w and the pH. By defining the pOH as the $-\log[OH^-]$, we can write the following expression:

$$pK_w = pH + pOH$$

$$14 = pH + pOH$$

This expression gives us the flexibility to calculate the pOH of a solution (when we are given the concentration of a base, such as NaOH) and translate it to the pH. Review Example 14.15 and Problems 77–80 for more work with pOH and pH.

8.3   Strengths of Acids and Bases

An important consideration of acid-base chemistry involves the strength of the compound. This relative strength of an acid or base can be described through the concept of electrolytes (See Chapter 14.7 of your text). For instance, the acid HCl is classified as a strong electrolyte, and hence, a strong acid. The acid will completely dissociate in water; one mole of HCl yields one mole of H_3O^+ and one mole of Cl⁻. The presence of these ions in water causes the solution to conduct electricity. Therefore, it is called an electrolyte.

Other compounds may dissociate only to a small degree. If a mole of acetic acid is added to water, only about 5% of the acid will dissociate. This limited dissociation does not allow enough ions to be present in the solution in order for the solution to conduct electricity. Hence, acetic acid is called a weak electrolyte.

When you are calculating the pH of an acid or base solution, you will need to determine whether or not the acid (or base) involved in the reaction is a strong or weak species. As mentioned in your text, the $[H_3O^+]$ for strong acids is usually the initial concentration of the reactants (i.e., HCl) because of complete dissociation, whereas the $[H_3O^+]$ for weak acids is determined from equilibrium calculations (as discussed in the next chapter of your text).

S E C T I O N 9

Skills for Chapter 15: Chemical Equilibrium

Goals: To calculate concentrations at equilibrium, and equilibrium constants.

Skills: Cross multiplication, powers, and roots.

C hapter 15 of your text introduces you to the concepts of chemical equilibrium. It is based upon having a known stoichiometric chemical reaction in which the equilibrium expression can be written. With this information, it is possible to predict the yield of a chemical reaction as well as determining the value of the equilibrium constant.

9.1 Calculations with Equilibrium Expressions

As an example of calculations you will use in your study of equilibrium, consider the chemical reaction A + B → 2C. The equilibrium expression for the reaction is:

$$K_{eq} = \frac{[C]^2}{[A][B]}$$

If you are given the equilibrium concentrations of A, B, and C, the equilibrium constant can be determined by entering these concentrations into the equation. For instance, if [A] = 0.10 M, [B] = 0.20 M, and [C] = 0.50 M, we can write:

$$K_{eq} = \frac{[0.50]^2}{[0.10][0.20]}$$

and

$$K_{eq} = 13$$

Examples 15.3, 15.11, and Skillbuilder 15.3 show the calculation of K_{eq} for a gas-phase reaction. Other problems concerning the calculation of K_{eq} are Problems 53-58.

A variation to this problem may involve the determination of [C] if [A], [B], and K_{eq} are given to you. If $K_{eq} = 13$ with [A] = 0.50 M and [B] = 0.25, let's arrange the above equation to solve for [C].

$$K_{eq} = \frac{[C]^2}{[0.50][0.25]}$$

Next, each side of the equation is multiplied by the denominator, [0.50][0.25]. This act leaves only the term $[C]^2$ on the right side of the equation.

$$13 \times ([0.50][0.25]) = [C]^2$$

By rearranging and multiplying the numbers we generate,

$$\sqrt{[C]^2} = \sqrt{1.6}$$

$$[C] = 1.3$$

The equilibrium concentration of C is 1.3 M. Compare this problem to Examples 15.4 and 15.12, Skillbuilder 15.4, and Problems 59-62.

There may be times in solving the equilibrium problems where the determination of a concentration involves taking the third or fourth root of the expression (See Problem 63). For instance, let's look at the production of ammonia gas using nitrogen and hydrogen gas:

$$N_{2(g)} + 3 H_{2(g)} \leftrightarrow 2 NH_{3(g)}$$

The equilibrium expression for this reaction is:

$$K_{eq} = \frac{[NH_3]^2}{[N_2][H_2]^3}$$

To solve for K_{eq}, we must raise the equilibrium concentration to the power of 3 for hydrogen, while squaring the NH_3 concentration.

You will need to use the powers function of y^x on your calculator (See Section 1 of this toolkit).

If, instead, you wish to solve for the hydrogen concentration using a given K_{eq} value and known equilibrium concentration of ammonia and nitrogen, the equation must be rearranged.

$$[H_2]^3 = \frac{[NH_3]^2}{K_{eq}[N_2]}$$

If K_{eq} = 9.6, [NH3] = 0.030 M, and [N2] = 0.050, what is [H2]? Plugging in the numbers, we can write:

$$[H_2]^3 = \frac{[0.030]^2}{9.6 \times [0.050]}$$

$$[H_2]^3 = 0.00188$$

$$[H_2] = (0.00188)^{\frac{1}{3}}$$

$$[H_2] = 0.123 \text{ M} \quad \text{using the } y^x \text{ function}$$

Important: In order to work these types of problems you must be able to write the correct equilibrium constant expression for the reaction. If you are in doubt as to how to construct the expression, look back at Chapter 15.4 and 15.5 of your textbook.

9.2   Calculations of Equilibrium Concentrations

A second use of the equilibrium expression involves the determination of the concentrations of the products and the reactants at equilibrium. If the stoichiometry of the reaction is known (allowing the equilibrium expression to be written) and K_{eq} is known, then the equilibrium concentrations may be found.

However, the solution to this problem is not always straightforward. In most cases, we only know the *initial* concentrations of the products and the reactants; we don't know the *equilibrium* concentrations. To calculate these equilibrium concentrations we must figure out how much the initial concentrations of products and reactants change, based upon the relationship defined by the equilibrium expression.

To help explain how such a solution to this problem is found, we will introduce here the "equilibrium checkbook". This teaching tool will allow us to quickly determine the concentrations of all equilibrium products and reactants. The checkbook concept is patterned very much like the checkbook that you are maintaining with a bank in your town. Just as keeping a good record of all deposits and drafts to the account enables you to balance the account, the equilibrium concentration of products and reactants can be readily determined with this strategy.

To see how the checkbook works, let's return to the "alphabet" reaction of Section 9.1 of this toolkit: A + B → 2 C. This balanced equation shows that 2 moles of product C are produced when 1 mole of A reacts with 1 mole of B. The equilibrium expression is then written as:

$$K_{eq} = \frac{|C|^2}{|A||B|}$$

For the purpose of this example, let's say that K_{eq} = 0.0100. The initial concentrations of both A and B are 0.500 M. The initial concentration of C is zero.

Because there is no C present, we can easily see that the reaction will proceed toward the right; that is, some C will be produced. However, there are times when you will be asked to perform calculations such as these when the concentration of the 'products' (those written on the right side of the equation) are not zero. To determine which direction a reaction will take (e.g., to the right or to the left), you should calculate the reaction quotient, Q. The formula seems identical to the equilibrium expression, but note that the initial concentrations, as noted by ( ) parenthesis, of the products and reactants are used rather than the equilibrium concentrations, as noted by [ ] brackets.

$$Q = \frac{(C)^2}{(A)(B)}$$

By comparing Q to K_{eq}, the direction of the reaction can be determined.

In the sample problem for demonstrating the equilibrium checkbook, we have the task of finding [C] from the equilibrium equation. The setup of the checkbook is shown below:

Species	A	B	2 C
initial			
Δ			
equilibrium			

In the top row (labeled species), the reactants and products of the chemical equation are written. The stoichiometry of the equations is also entered in the top row of each column. Note that the double line separates the reactants and products and serves in place of the ↔ symbol. The other rows are labeled: initial, Δ, and equilibrium. These entries will pertain to changes in the concentrations of the reactants and products during the course of the reaction.

With the given information, we can set up the entries in the equilibrium checkbook. Our initial concentrations are entered in the initial row under the proper column of each reactant and product.

Species	A	B	2 C
initial	0.500 M	0.500 M	0 M
Δ			
equilibrium			

The next stage is to determine the changes (withdrawals and deposits) to each account. From the stoichiometry of the reaction, every 1 mol of A and 1 mol of B that react produce 2 moles of C. With this reasoning, if x mol of A and x mol of B react, then $2x$ moles of C are produced. Since A and B are reactants, their concentrations should be depleted as the reaction proceeds. This "withdrawal" can be entered in the checkbook in the Δ row as $-x$. Similarly, the "deposits" of $+2x$ to the C column can be entered, as shown below:

Species	A	B	2 C
initial	0.500 M	0.500 M	0 M
Δ	$-x$	$-x$	$+2x$
equilibrium			

The last step is to add the activity in each account. This "bottom line" is the equilibrium concentration as a function of initial concentrations and x. By adding the initial and Δ entries together for each product and reactant we find:

Species	A	B	2 C
initial	0.500 M	0.500 M	0 M
Δ	$-x$	$-x$	$+2x$
equilibrium	0.500 M $- x$	0.500 M $- x$	2x

The entries in the bottom row are the equilibrium concentrations as a function of the initial concentration and x. Since we want to know the equilibrium concentrations as a single number rather than in terms of initial concentrations and x, we must solve for the value of x. Once we know x, we can directly express [A], [B], and [C].

The solution lies in the use of the equilibrium expression. By substituting the data into the equilibrium expression, we can write:

$$0.0100 = \frac{[2x]^2}{[0.500 - x][0.500 - x]}$$

Note that we have one equation and one unknown. To solve for x, we must employ our algebraic skills. For this problem, the solution lies in noting that the equation can be simplified to:

$$0.0100 = \frac{[2x]^2}{[0.500 - x]^2}$$

If we now take the square root of both sides, the x^2 term can be reduced to x, thereby making the equation much easier to solve.

$$\sqrt{0.0100} = \sqrt{\frac{[2x]^2}{[0.500 - x]^2}}$$

which simplifies to:

$$0.100 = \frac{[2x]}{[0.500 - x]}$$

By cross multiplication and rearrangement, the following equations show the solution for x:

$$0.100 \times [0.500 - x] = [2x]$$

$$0.0500 - 0.100x = 2x$$

$$0.0500 = 2.100x$$

$$x = \frac{0.0500}{2.100}$$

$$x = 0.024$$

Having found x, the equilibrium values of A, B, and C can be calculated.

$$[A] = (0.500 - 0.024) = 0.476 \text{ mol/L remaining of A}$$

$$[B] = (0.500 - 0.024) = 0.476 \text{ mol/L remaining of B}$$

$$[C] = (2 \times 0.024) = 0.048 \text{ mol/L produced of C}$$

These values are the equilibrium concentrations of the reactants and products.

The use of the checkbook will allow you to set up and solve every equilibrium problem that is discussed in your text. Not all problems will be solved by the same exact algebraic methods shown above, but the solutions are similar. Try this approach on Problem 64.

9.3 | Solubility

Another form of equilibria discussed in Chapter 15.11 is solubility. The equilibrium constant that describes this process is called K_{sp}, the solubility product. Consider the process of dissolving AgCl in water.

$$AgCl_{(s)} \leftrightarrow Ag^+(aq) + Cl^-(aq) \qquad K_{sp} = 1.8 \times 10^{-10}$$

Since this is a heterogeneous reaction, the equilibrium expression is written as:

$$K_{sp} = [Ag^+][Cl^-]$$

The checkbook method described in the previous section of this toolkit is helpful in determining the molar solubility of AgCl.

If excess AgCl is placed in water, the K_{sp} expression defines the maximum allowable product of two species (that is, the product of the soluble ions). Setting up the checkbook, we can write

Species	Ag^+	Cl^-
initial	0 M	0 M
Δ		
equilibrium		

Because $AgCl_{(s)}$ is not part of the heterogeneous equilibrium expression, it is not listed here. (Remember, the concentration of a solid cannot change, therefore only the soluble species in this reaction are listed.)

The change column is determined from the stoichiometry of the equation. Here 1 mole of $AgCl_{(s)}$ yields 1 mole of $Ag^+(aq)$ and 1 mole of $Cl^-(aq)$. If x moles of AgCl dissolve, then the checkbook becomes:

Species	Ag$^+$	Cl$^-$
initial	0 M	0 M
Δ	+x	+x
equilibrium		

The equilibrium concentrations are calculated as:

Species	Ag$^+$	Cl$^-$
initial	0 M	0 M
Δ	+x	+x
equilibrium	+x	+x

If we substitute these values for the equilibrium concentrations into the equilibrium equation, we obtain:

$$K_{sp} = [x][x]$$

$$K_{sp} = [x]^2$$

$$x = \sqrt{K_{sp}}$$

Solving for x will tell us the molar concentrations of each ion in solution as well as the molar amount of solid that was dissolved. Plugging in K_{sp} and solving this expression yields:

$$x = \sqrt{1.8 \times 10^{-10}}$$

$$x = 1.4 \times 10^{-5}$$

The molar solubility of AgCl$_{(s)}$ is 1.4 × 10^{-5} M. Compare this to Examples 15.9 and 15.15, and Skillbuilder 15.9. Also, try Problems 81-84 with this approach.

SECTION 10

Skills for Chapter 17: Radioactivity and Nuclear Chemistry

Goals: To determine the half-life of a radioactive element.

Skills: Multiplication, division, roots, and scientific notation.

In this chapter you will be introduced to problems that concern the decay of radioactive elements. The primary calculation of these examples concerns the half-life of a radioactive element. This is described in detail in Chapter 17.5 of your textbook and is shown in Example 17.4 and Skillbuilder 17.4

The mathematics of this example involves multiplying the original amount of a radionuclide by a fraction in order to find the amount of undecayed radionuclide. This fraction is defined as $1/2^n$, where n is the number of half-lives involved in the decay. It is written as:

$$\text{amount undecayed} = \text{original amount} \times \frac{1}{2^n}$$

Problem 73 asks us to determine the mass of a radioactive nuclide that would remain after a period of 10 days. The problem states that the half-life of the nuclide is 2 days and the sample originally contains 1000 atoms. The decay time of this problem is 5 half-lives. By setting n=5, the fraction is 1/32, or 0.03125. Multiplying this fraction by the original amount of nuclide (1000 atoms) tells us that 31 nuclides would remain after 10 days. Try Problem 74 is with this approach.

A second use of this equation allows us to estimate the time required for a radioactive compound to decay to a prescribed level. For instance if the half-life of Bi-214 is 19.7 min, how long will it take for only 25% of the material to remain? If the original amount of Bi-214 was 1.00g, we are trying to determine when only 0.25 g will remain. Using the equation above, we write:

$$0.250 \text{ g} = 1.000 \text{ g} \times \frac{1}{2^n}$$

$$\frac{0.250 \text{ g}}{1.000 \text{ g}} = \frac{1}{2^n} \qquad \text{dividing terms}$$

$$0.25 = \frac{1}{2^n} \qquad \text{simplifying}$$

$$\frac{1}{0.25} = 2^n \qquad \text{inverting}$$

$$4 = 2^n \qquad \text{simplifying}$$

$$n = 2 \ \text{ half-lives}$$

This time period would be 2×19.7 min, or 39.4 min for this amount of decay to occur. After only 39.4 minutes the amount of Bi-214 would be 0.250 g or less. Compare this example to Example 17.4 and Problems 75-80.

SECTION 11

Listed below are a series of questions designed to check your math skills. Please take no more than 30 minutes in answering them. You will find the answers in the back of this booklet.

1. How many significant figures are in the number 0.101?

2. Solve for x in the following equation with care to express the correct number of significant figures:

$$x \;=\; 12.011 + 1.00797$$

3. Solve for x in the following equation with care to express the correct number of significant figures:

$$x \;=\; \frac{1.057}{10.3}$$

4. Express the number 0.000356 in scientific notation.

5. Determine the mean, $\bar{x}$, for the following data: 25.12, 25.29, 24.95, 25.05 and 25.55.

6. Solve the equation: $x = 2.0 (\sin 35°)$.

7. Solve the equation: $0.405 = \cos(x)$.

8. Determine the value of x in the following equation:

$$(20.3) \times (0.0171) = \left(\frac{x}{60.3}\right) \times (1.35 \times 10^3)$$

9. What percentage of 15.5 is 1.25?

10. Solve the following equation for x:

$$\frac{1}{x} = \frac{4.3}{6.8}$$

11. Solve the following equation for x:

$$\frac{0.57}{0.22} = \frac{1.2}{x^2}$$

12. Calculate the value of x, where $x = (6.23 \times 1.22)^3$.

13. What is the fourth root of 17?

14. Solve the following equation for x:

$$x = 8.3 \times \left(\frac{1}{2^2} - \frac{1}{4^2} \right)$$

15. Determine the value of x in the following equation:

$$2.52 = \sqrt{\frac{1}{x}}$$

16. Calculate the value of x, where $x = \log (0.017)$.

17. Find the value of x, where $x = e^{\ln(12)}$.

18. Determine the value of x, where $1.53 = \log (x)$.

19. Solve for x in the following equation:

$$x = \frac{-1}{0.030} \times \ln\left(\frac{0.20}{0.50} \right)$$

20. Solve for x in the following equation:

$$x = \left(1.00 \times 10^3 \right) e^{\left(\frac{-0.0220}{1.33} \right)}$$

SECTION 12

Chemistry and Writing

12.1 The Scientific Notebook

The scientific notebook is the scientist's own record of experiments performed and phenomena observed. Beginning with the first student laboratory report, there are special requirements for recording experimental results. The requirements may seem rigid at first, but they are very understandable in the light of the purposes of the notebook.

For the professional scientist, the claim to original work is found in the scientific notebook. Millions of dollars in patent rights may depend on the existence of a properly dated and authenticated scientific notebook. Many of the rules that are followed in recording data follow from this important function of the notebook. Nothing is ever erased; an incorrect reading is crossed out, and the correct one written beside it. Work is recorded in a bound notebook with pages that cannot be removed or added. Every entry is dated, signed, and countersigned by the scientist in charge of the laboratory. All these rules are designed to produce a record that will constitute proof not only of what experiments were performed, but of the exact date. This is important, because if two scientists make the same discovery, the first one to do so will gain all the legal rights and most of the credit for the work. Obviously, it is more important to have a complete and original record than a perfectly neat one. A few crossed-out readings are not uncommon, and a few blots from spilled chemicals are not unheard of either. These are preferable in the laboratory notebook to a perfect page that has been copied over at a later date and no longer constitutes an authentic original record. **Under no circumstances is data to be recorded on loose paper rather than directly into the notebook!**

Another important function of the notebook is to record the procedure and observations so clearly and completely that the experiment can easily be repeated at a later date. Experiments that cannot be repeated by the same researcher or by other laboratories are soon discredited. For the student in the laboratory, complete notes are important as well. If something goes wrong, it should be possible to find the error in procedure from the experiment notes. At times, the numbers in the crossed-out data entries tell an interesting story. Occasionally an interesting and unexpected phenomenon will be observed that merits further study. Always, a complete, clear record of what has happened in the laboratory is essential.

In order to be a complete record, each experiment entered into the notebook should include certain features. The scientist's **name** and the **date** should always be entered. The **title of the experiment** being performed is an important element that is often neglected. "Chemistry Lab" is an inadequate substitute for the experiment title, which is usually readily available. Often it is useful to begin by writing the **objective**, or the purpose, of the experiment. Stating the objective clearly helps both the experimenter and the reader of the notebook to understand the experiment. A complete record of experimental **procedure** is essential, either as a step-by-step description or by a complete reference to a standard experimental procedure. If a standard procedure is given, great care must be made to note any deviations from that procedure. A list of **materials and equipment** used can be a great help in organization if it is included as a part of the experimental procedure.

Though the laboratory notebook does not have to be perfectly pristine, it is certainly desirable that it should be as organized as possible. Some time and thought spent in planning before the laboratory period begins will result in a better notebook and a more successful experiment. **The date, title, experimenter's name, and objective of the experiment should be entered before the experiment begins.** If the experimental procedure that has been provided does not already give **labeled data tables** for an experiment, it is worth some time and thought to set up such tables before entering the laboratory, rather than waste time during the experiment deciding how to do so. Ample space should be provided not only for the expected data, but also for corrections and notes. Unused space can be crossed out later as necessary, though extra pages are never torn out. Sometimes only the right-hand pages of the notebook are used, leaving the other pages free for later notes or calculations. Individual research laboratories or student laboratories may have standard notebooks or forms in which to write laboratory results. All of them share the basic objective of recording in a useful way the scientist's actions, observations, and thoughts while in the laboratory.

12.2 The Scientific Report

When the scientist makes a formal written report of experiments performed in the laboratory, the report follows a generally accepted outline. Introduction, results and discussion, and conclusions follow in order as separate sections and are clearly labeled. Lists of references and even the title are treated in standard ways.

The **title** of a scientific paper is seldom an occasion for creativity. Titles for articles in scientific journals are carefully constructed from words that will be useful key words for information searches by computer. Titles for student laboratory reports are usually indicated in the assignment. As with the laboratory notebook, "Chemistry Laboratory" is unacceptably vague as a laboratory report title. Abbreviations as part of a title should be avoided.

The **Introduction** section should make clear to the reader the purpose and the background of the experiment. The objective of the work that is being discussed should always be clearly stated. It may be appropriate to discuss the basis of the experimental methods that were used as well as the scientific theory on which the work is based. Usually a well-written introduction makes use of written resources in the form of scientific books and papers, which must be listed in the references cited and footnoted with the appropriate reference.

The **Experimental Procedure** section explains in detail exactly how the experiment was conducted. It should be possible to reproduce the experiment using the information found in this section. If standard procedures are used and not explained in detail, a reference should be given. A list of materials and equipment is often a useful component in this section. It includes all chemicals used, including the concentrations of solutions, and all special equipment.

The **Results and Discussion** section includes the data that were obtained in the experiment together with an explanation of the data. Often it is useful to organize the results of the experiment into tables, and sometimes graphs are required as well. All tables and figures should be titled and numbered. All columns in tables and both axes of a graph should be carefully labeled, not omitting units. If calculations have been performed, the equations used should be clearly indicated and enough information about the calculations should be included so that they can be clearly followed. The precision and accuracy of the results should be calculated by standard statistical methods if appropriate to the experiment.

The **Conclusions** section contains the thoughts of the experimenter about the significance of the work performed. Each part of the experiment should be discussed. Numerical results should be evaluated, and the meaning of any statistical calculations explained. The success of the experiment should be evaluated by referring to the objective of the experiment as presented in the introduction. Was the experiment successful? Were the objectives met? What is the overall significance of the experiment?

The **Literature Cited** section lists all the references used in preparing the report. This section is the most formalized of all in its format. Each scientific journal has a slightly different style which contributors must follow to the letter. Student reports may also be required to follow a certain form. The best way to write this section is with the help of an example. Often college courses use scientific journals as models. The *Journal of Chemical Education*, *Analytical Chemistry*, and the *Journal of the American Chemical Society* are examples of chemical journals which have been used in this way; the *Journal of Organic Chemistry* is often used in organic chemistry courses. When giving references it is important to notice carefully all words that are set in italics or boldface in the example references. Typesetters use different fonts for italics and boldface that are difficult to reproduce when typing or handwriting, though many word processing programs are able to reproduce them. Words that are set in **italics can be indicated by an underline. Boldface can be represented by a wavy underline.** Typically, a reference to a book appears as follows:

REID, R. C.; SHERWOOD, T. K.; PRAUSNITZ, J. M. *Properties of Gases and Liquids*; McGraw-Hill: New York, 1977.

A reference to a scientific journal follows this general form:

LEE, L. G.; WHITESIDES, G. M. *J. Am. Chem. Soc.* **1985**, *107*, 6999.

12.3   Technical Writing

Scientific writing is not limited to scientific journal articles. Scientists on every level are more likely to achieve success if they are able to describe their work and explain its significance to others. Technical writing can vary from a brief explanation of how to use a piece of equipment to a lengthy report on the activities in the laboratory. Technical writers produce articles written for the layman explaining technical subjects in understandable terms. Effective technical writing is a job skill that is very much in demand. College-level assignments that involve report writing on technical subjects require the same considerations as professional writing.

First, consider the audience. Will a skilled professional or a layman read the material? If it cannot be assumed that the reader is familiar with the basic principles of the field being discussed, then the writing must include some basic background information, with special attention given to explaining technical vocabulary that may not be understood by the reader.

Most writing projects begin with a visit to the library to find appropriate source materials. Again, the level of the project will determine how the literature search is conducted. The original research reports contained in scientific journals can be found through indexes such as those provided in *Chemical Abstracts*. Using the abstract indexes is a skill that must be developed through practice. Chemistry students usually are given a special course in the chemical literature that includes training in the use of *Chemical Abstracts*. Many science reports, however, require only limited use of original research papers. Science encyclopedias and dictionaries, along with periodicals written for the layman, can provide the background information for a science report and may indicate as well the authors and topics that might be explored in a more detailed technical search. Science and technology encyclopedias useful as sources of background information include:

Harper Encyclopedia of Science
McGraw-Hill Encyclopedia of Science and Technology
Van Nostrand's Scientific Encyclopedia

More specialized encyclopedias include:

Encyclopedia of Chemical Technology
Encyclopedia of Physics
McGraw-Hill Encyclopedia of Energy

Dictionaries can be useful in defining technical terms and concepts. Those that are useful in chemistry-related topics include:

Chamber's Dictionary of Science and Technology (McGraw-Hill)
Chemist's Dictionary (Van Nostrand)
Hanckh's Chemical Dictionary (McGraw-Hill)
McGraw-Hill Dictionary of Scientific and Technical Terms

Facts and data can be found through the many scientific handbooks. Some of the handbooks used in researching chemistry papers include:

CRC Handbook of Biochemistry
CRC Handbook of Chemistry and Physics
CRC Handbook of Environmental Control
Merck Index

Review articles in periodicals such as *Scientific American* give useful information on a variety of scientific topics. They can be conveniently found through the *General Science Index*, which provides a comprehensive subject index to English language periodical literature in the sciences. On-line computer search services are increasingly used to locate periodical references. A major resource of the library not to be neglected is the expertise of a good science librarian.

Technical writing depends no less than any other form of writing on the basic language skills of the writer. Incorrect spelling and grammar can mar the effect of the most interesting and original narrative. A good guide to English usage belongs next to a dictionary on the writer's desk. Good writing style is developed through practice in writing and rewriting. A clear, direct style contains no unnecessary words. Consider the following example:

```
At this point in the experiment the mixture was
heated up through the use of a hotplate.
```

A much-improved version is:

```
The mixture was heated with a hotplate.
```

Some science publications prefer that use of the first person ("I heated the mixture") be avoided. Use of the passive voice "the mixture was heated" is then indicated. In other uses the more direct form of the active voice may be preferred, as in "We decided to heat the mixture" rather than "It was decided that the mixture should be heated." When writing instructions, the imperative is often a good choice: "Heat the mixture on a hot plate" is more direct than "The mixture should be heated on a hot plate."

There are many references available to you to help you develop the valuable skill of communicating information. General references include:

W. STRUNK, Jr.; E. B. WHITE. *The Elements of Style*, MacMillan: New York, 1979.

MARGARET SHERTZER. *The Elements of Grammar*, MacMillan: New York, 1986.

References pertaining to technical information are:

B. EDWARD CAIN. *The Basics of Technical Communicating*; American Chemical Society: Washington, D.C., 1988.

ANNE EISENBERG. *Writing Well for the Technical Professions*; Harper and Row: New York, 1989.

12.4 A Notebook Example

The following three pages illustrate the proper manner in which a laboratory notebook should be kept. These pages are reproductions of a student's analytical chemistry laboratory notebook. The style of this notebook conforms to the guidelines presented in Section 12.1 of this booklet.

Before entering the laboratory the student has written the introduction and experimental section. A section of data/results was begun as the student collected information. (Note that the student recorded the weights of several tablets as well as the time required to complete the coulometric titration of the ascorbic acid tablets.) This entry is reproduced on page 71.

Next, the student was required to perform several calculations with these data. They were begun in the laboratory and are found on the next page (p. 72). (Note that this page is the reverse page of the data entry. It is the left-hand page of the notebook.) Normally, the right-hand side of the notebook is used for data entry and analysis, while the left-hand side is used for "scratch" calculations. This method eliminates the need for "loose" paper for initial calculations and other scribbles.

The last page (p. 73) contains the conclusions. Note that the student corrected the calculated values and his grammar by drawing a line through the unwanted entry. Erasure or correction fluid was not used.

This simple system of using the different sides of the notebook for formal entries and simple calculations can greatly enhance your ability to organize your data and to present your findings.

Coulometric Titration of Ascorbic Acid with Iodine   3/23/00

Introduction · The purpose of this experiment is to determine the weight
          percent of ascorbic acid in a vitamin tablet.

Experimental : Add 0.1M KI solution and a spatula of soluble starch.
          Cautiously add 1sec increments to the current until a pale
          blue color appears This will be the matching color.
          Crush a vitamin tablet and weigh 20 to 30mg to the
          nearest 0.1mg; transfer to the cell. Turn on current.
          Repeat the procedure with 2 more samples from the
          tablet. Continue to analyze other tablets.

Data/Results: multiplier 0.5 0.05 = 48.25mA
          250mg tablets / 361mg tablet #1, 364.8 mg tablet #2
          sample   1   0.0232g tablet   389.2 sec
                   2   0.0225g   368.4 sec
                   3   0.0243g   ———
                   4   0.0249g   385.0
                   5   0.0240g   359.1
                   6   0.0234g   353.0
                   7   0.0240g   366.3

Actual Weight	Time	AB Ascorbic Acid		Weight %
0.0232g	389.2sec	0.0171g	Tablet 1	73.7
0.0225g	368.4sec	0.0162g		72.0
0.0249g	385.0sec	0.0170g		68.3
0.0240g	359.1sec	0.0158g	Tablet 2	65.8
0.0234g	353.0sec	0.0155g		66.2
0.0240g	366.3sec	0.0161g		67.1

          Tablet #1 contained on average 263.0 ± 4.3mg ascorbic acid.
          Tablet #2 contained on average 244., ± 4.2mg ascorbic acid.

263.0 ± 4.3

244.1 ± 4.2

$95\% \; CL = \bar{x} \pm \dfrac{ts}{\sqrt{N}}$

$= 263.0 \pm \dfrac{12.7(4.3)}{\sqrt{2}}$

$= 263 \pm 38$

$= 244.1 \pm \dfrac{3.18(4.2)}{\sqrt{2}\,4}$

$= 244.1 \pm 9.4$

$ 6.7$

$\bar{x} - \bar{\mu} = 263 - 250$

$\phantom{\bar{x} - \bar{\mu}} = +13$

$\dfrac{ts}{\sqrt{N}} = \dfrac{(12.7)(4.3)}{\sqrt{2}}$

$\phantom{\dfrac{ts}{\sqrt{N}}} = \pm 38$

$\bar{x} - \bar{\mu} = 244 - 250$

$\phantom{\bar{x} - \bar{\mu}} = 6$

$\dfrac{ts}{\sqrt{N}} = 6.7$

Conclusion: The value given on the box for each ascorbic acid tablet was 250mg. ~~We~~ The obtained results for two tablets were 263.0 ± 4.3 mg and $244., \pm 4.2$ mg. The results are close to the reported value. Some error can be expected in this experiment because the calculations depend on an observed endpoint, and although it is tried, it is sometimes hard to pick the same endpoint each time based on color.

The third run on the first tablet was discarded because the electrode was not in the solution and therefore no reaction was taking place.

The 95% CI produces results of 263 ± 38 for the first tablet, which rounds to $263 \pm 3_8$. This is mainly high because it is based on two readings. The second tablet gave a result of $244., \pm 4_{7}$ mg. This is based on four readings.

A t-test on the first tablet gives $+38 < +13$, which says that our result is good and would, on average, be correct 95 of 100 times. Our second tablet gives $-6.7 < 6$, which suggests that our second value can also be accepted.

The overall reliability of Coulometric titrations is good as long as you pick an endpoint and can reproduce it.

SECTION 13

Careers in Chemistry

Chemists often find themselves in careers that seem unrelated to traditional chemistry, yet draw heavily on chemical knowledge and skills. Some chemists go into management; management positions in science-related industries increased by 30% between the years 1994 and 2000. Technical sales, patent law, and forensic science are only a few of the careers open to holders of chemistry degrees. Laboratory workers are needed in materials science, polymers, and biotechnology; all these fields depend on the molecular science of chemistry. The traditional skills of chemical analysis and synthesis are in demand for such areas as environmental testing and pharmaceutical development. A wide variety of rewarding career options is available for those with chemical training. The overall unemployment rate for chemists is very low, averaging about 1%.

13.1 Materials Science

One of the major predicted growth areas in the economy is in the area of materials science. The new superconducting materials that promise major breakthrough applications in fields as diverse as communications and transportation are the products of materials science. The development of new graphite materials is resulting in new types of tennis rackets, aircraft, and auto bodies. New ceramic materials are being researched for a variety of uses, including automotive pollution control. Fiber optic materials are revolutionizing communications.

Like many of the fields that offer exciting new job opportunities, materials science is interdisciplinary, requiring knowledge of chemistry and physics. The field employs both chemists and chemical engineers. Courses in polymer science, metallurgy, and computer science are also helpful. Industry employs the largest number of materials scientists; while government and university positions are possible in this field as well.

13.2 Polymer Science

The long-chain giant molecules called polymers form a special class of materials that find a wide variety of uses. Synthetic fibers such as nylons and polyesters are the basis of a major industry. Packaging materials as diverse as Saran Wrap and Styrofoam are made of polymers. Televisions, computers, toys, tapes, and CDs all make use of polymeric materials. Increasing concern about solid waste disposal is prompting research on plastic recycling and degradability. A promising research area involves modification of polymer properties to make materials that will be compatible with human tissues for medical transplants. Preparation for an industrial career in polymer science may involve a degree in chemistry, chemical engineering, or polymer science and engineering.

13.3 Environmental Science and Technology

Increasing concern about the environment has created numerous job openings for scientists who want to find solutions to the problems caused by pollution. Understanding the chemical and biochemical reactions that produce and consume chemicals like carbon dioxide and methane in the atmosphere is critical to understanding the possible long-range warming process known as the greenhouse effect. Chemists discovered that the chemical reactions of CFCs, or compounds containing carbon, fluorine, and chlorine, with ozone in the upper atmosphere were causing ozone depletion and a resultant increase in the amount of harmful ultraviolet radiation that is reaching the Earth's surface. Now chemists must discover substitutes for these compounds that will replace them in such uses as refrigerants for refrigerators and air conditioners. Automobiles and factories must be designed so that their combustion processes do not foul the air. Factory effluent must be treated or recycled so that streams and groundwater are not polluted.

State and federal agencies such as the Environmental Protection Agency employ chemists to monitor pollution and help to find ways to decrease its sources. Waste management companies recycle materials or dispose of them responsibly, an increasingly technical task. Chemists, biochemists, and chemical engineers all find employment in areas of environmental technology.

13.4   Biochemistry and Biotechnology

Biochemists study the chemistry of living systems. Understanding the reactions which occur in the human body and in animals, plants, insects, viruses, and microorganisms makes new approaches to curing disease and improving food technology possible. New information from the human genome project will have profound implications for our society. Colleges and universities employ almost half the biochemistry workforce; the rest are employed by government agencies or private companies.

Biotechnology is a burgeoning new interdisciplinary field employing biochemists, chemists, and biologists. Most of these are employed in industry, either by established pharmaceutical and agricultural companies or by new biotechnology venture firms in this rapidly growing area. Through biotechnology, scientific breakthroughs in molecular biology are used to develop new products for the commercial marketplace. Microorganisms able to produce insulin or the human growth hormone are being produced on a large scale. Specific microorganisms are being produced with the goal of consuming oil spills or hazardous chemical waste. Genetic variation of plants may produce varieties which are richer in nutrients or resistant to insects. These are but the beginnings of the commercial applications which may be expected in biotechnology.

13.5 Medicinal and Clinical Chemistry

Biochemistry and biotechnology are improving both our understanding of diseases and the methods of treatment, but they are not the only fields of chemistry related to medicine. Medicinal chemistry and clinical chemistry are specialties in which chemists have a direct impact on health care.

The medicinal chemist develops new therapeutic agents. Chemical compounds are designed and synthesized with the aim of producing molecules, which will act on one area of the body, such as a certain part of the brain. Once the compound has been synthesized, it is ready for the long series of tests for both positive and harmful effects, which all new drugs must undergo. Tests may indicate that further modification of the molecule's structure is required. To prepare for this work, which links chemistry, biology, and medicine, requires training in chemistry, pharmacology, or medicinal chemistry.

The clinical chemist applies the techniques of analytical chemistry to body fluids. A variety of chemical instrumentation is used, and the modern clinical laboratory makes extensive use of computer-automated testing. Administrative duties may form an important part of the clinical chemist's work. Clinical tests are performed both to monitor normal body functions and to test for therapeutic and toxic drug levels in the body. Often critical decisions about health care are based on laboratory results. Many clinical laboratories are involved in research-related testing, to find the results of new procedures, drugs, and equipment or to support basic research. Clinical chemists are employed by hospitals, universities, government and industry.

13.6 Forensic Science

Forensic chemists (also called Crime Scene Investigators, or CSIs) apply the skills of analytical chemistry in the crime laboratory. Suspected samples of drugs must by analyzed for authenticity by chemical instrumentation if they are used as legal evidence. Body fluids and samples of body tissue are provided by the medical examiner for analysis in the case of a homicide, and cases of sexual assault involve laboratory testing as well. Traces of blood or of gunshot residue can be revealed by chemical tests. Other skills are required of the general forensic scientist, such as investigation of the crime scene. Though these are not directly related to chemistry, some of the most eminent forensic scientists were trained as chemists. Most forensic scientists are employed by government laboratories.

13.7 Radiochemistry

An exciting area of chemistry which is experiencing a severe shortage of trained workers is radiochemistry. This specialty is needed in a multitude of application areas, some of which are growing rapidly.

Nuclear medicine is seeing rapid advances in such areas as the use of monoclonal antibodies in cancer research. The radiopharmaceutical industry needs radiochemists both to manufacture radionuclides and to oversee waste disposal and protection from radiation hazard. Positron emission tomography (PET) is a diagnostic application of radiopharmacology, which requires radiochemists to staff PET centers and to develop and provide radionuclides and labeled compounds. PET is expected to expand rapidly as its use shifts from primarily research use to more routine medical applications.

Nuclear power production requires radiochemists both in power plants and in related industries to oversee the running of plants and to supervise waste treatment.

Environmental chemistry makes use of radiochemistry in several ways. Accidental emissions from power plants such as the Chernobyl accident require skilled monitoring to assess environmental and health effects. Radionuclides can be used as tags to monitor complex systems such as the buildup of greenhouse gases or the sources of atmospheric CFCs, which destroy the ozone layer. Neutron activation analysis can analyze for trace elements in rocks and help explain Earth's history.

13.8 Patent Attorneys and Patent Agents

A background in chemistry can serve as preparation for the career of patent attorney or patent agent. A patent attorney must have a law degree as well as a science degree. A patent agent may have a degree in chemistry, physics, engineering, or related technological fields. For certification, the patent agent must pass an examination on patent procedures and rules. The patent agent examines the patent literature to determine whether a client's invention is patentable and writes patent applications. Patent lawyers may also represent a client in litigation on licensing, trademark or copyright issues, trade secrets cases, and antitrust cases. Patent agents and attorneys are always well paid and in short supply. Their work involves communication skills as well as scientific training. They are in demand both by government offices and private law firms.

13.9 Chemical Education

Chemistry is taught at both the high school and the college level. College chemistry is taught at two-year and four-year colleges and at universities. Teaching positions at larger universities are likely to involve greater emphasis on graduate education and research than on undergraduate instruction. Demand for high-school teachers is increasing in most areas of the country, especially with the recent passage of No Child Left Behind and the new requirements of high school science teachers.

Teachers at all levels speak of the satisfaction involved in interacting with young minds and influencing future careers. Individual freedom is often greater than in many other types of employment, and scheduling is more flexible. College professors have many obligations in addition to teaching, however. Committee responsibilities are required as a part of participation in college governance, and administrative duties may be part of some professors' workloads. Research involves not only study and planning but also the training of graduate students and postdoctoral fellows and the writing of proposals for research funding. Financial rewards may not be as great as those of the industrial chemist, especially at the high-school level and at smaller higher education institutions. For those who teach, the quality of life in the work experience is often a major part of the reward.

13.10 Chemical Information Careers

Chemical information is produced at an increasingly rapid rate as new discoveries are made and the results published in scientific journals. Several career possibilities are available to the chemist who can help others to locate and understand the relevant chemical information for their needs.

The chemical information specialist is familiar with both the chemical literature and the on-line computer services through which chemical information is increasingly accessed. Libraries use chemical information specialists, but chemical companies need them also to help their researchers. Often toxicological and environmental information is needed as well as experimental data retrieval. This special kind of problem-solving expertise should remain in demand as information retrieval becomes an increasingly complex field.

Abstracting services summarize and index chemical discoveries so they can be found by their researchers. They employ chemists to prepare information both for written abstract material and for on-line search services.

Science writers are needed on all levels, both to edit and write scientific publications for the scientists and to interpret scientific information in terms the general public can understand. Research organizations, medical centers, technical companies, and government agencies all need writers who can communicate clearly with a variety of audiences. In addition to print media, radio and television employ experts in science communication.

13.11 Other Chemistry Careers

This list of some current options in chemistry careers is by no means complete. Chemists work in museums and metallurgy plants, in food science laboratories and consumer testing services, in adhesives research and photography laboratories. The materials of our modern world are developed by chemists, and chemists monitor the safety of these products and of our environment. The materials of our future and discoveries that will prolong our lives and improve the quality of life will be the products of chemical research.

SECTION 14

Study Aids for Chemistry

Chemistry is a field of science whose basic foundations and concepts go back for many centuries, if not a millennium. All of this activity by scientists has produced a great body of knowledge, which is collectively called "chemistry." The vastness of this body of knowledge is more than one individual could ever know. Even with a lifetime of study, no one person could ever be well learned in all the areas of chemistry.

However, the study that you are to embark on this year uses a text, **Introductory Chemistry**, which is a well-organized guide into the basic concepts of chemistry and into the connections that chemistry has to every facet of modern life. Over the coming year, you will be exposed to discoveries and wonders in chemistry that now define our modern life and offer ways to improve the quality of life for all people.

Having said this, let us emphasize that the material presented in this course will not require all of your waking hours to successfully comprehend and master. You have probably heard "horror stories" from older students concerning chemistry. As a chemistry teacher (who has taught nearly eight thousand students over two decades), let me say that their stories are quite overrated. It has been my perception that the students who do not do well in chemistry are often the ones who have not learned how to study effectively.

The habit of good studying is not one that comes naturally. It is a learned process. It is not a hard process; in fact, most of the basis of these habits is founded in good "common sense."

On the next few pages, you will find suggestions on how to develop good study habits. Also listed are suggested ways to get the most out of your textbook, ways to relax while studying chemistry, and tips on taking exams. You may already be practicing many of these techniques. If you see new ones, try to incorporate them into your study routine.

Life is a learning process. While the focus of your learning will change as you go from school to a job, the habit of good studying really does not change. Good study habits can make you a "quick learner."

Most of all, good study habits can help you enjoy chemistry. That's right, enjoy chemistry! It is truly filled with wonder and amazement. It is rich in history and is relevant to your everyday life. The understanding of the concepts presented in your text and reinforced in this booklet will most certainly allow you to see the connection of chemistry to all that is around you. While good study habits will allow you to go beyond the memorization of the material and into critical thinking, it is the skill of critical thinking that sets the stage for your enjoyment of chemistry. This enjoyment from your understanding of chemistry can be truly wondrous.

14.1   Good Study Habits

Good study habits are not among our "basic instincts." Most of us have to work long and hard to develop good study habits. Here are some concrete things you can do to promote your success as a student.

- Set a long-term goal, that is, a career goal. Goals keep us going when the going gets rough.
- Set short-term goals, that is, daily or weekly goals. Short-term goals motivate us and keep us on target.
- Make a realistic daily, weekly, and term-long schedule. Post copies of your daily schedule where others can see it. That will force you to stick to it. Also schedule relaxation time for yourself! We all need that!
- Observe "good students," and make yourself look and act like a good student. Soon you will start thinking of yourself as one. We often have to "act" ourselves into a different way of thinking.
- Create a study area that meets your needs. The study area should have few distractions and be quiet. Try to keep it uncluttered. Having "a place" to study helps you get in the mood to study.

- Select the best time for you to study. We all have different "internal clocks." Study when you are most alert.
- Reflect and review daily. *Reading a lesson and doing problems is not studying.* You read the lesson and do the problems to gain the knowledge you need to study the material. Studying involves moving material from memory to understanding.
- If you can't explain a concept in your own words and relate it to something you already know, you probably don't understand it and won't remember it very long.
- Get to know others in your class, and form a serious study group. Then study together at least once a week. It will keep you "honest." It is easy to slip into sloppy study habits and fool yourself into thinking you know more than you do. Having to explain ideas in a study group will bring you back to reality very quickly. We learn and learn how to learn from working with others.
- Work to stay relaxed while you are studying.

14.2 Reading Your Textbook

If your assignment is to read pages 88 to 108, the worst thing you can possibly do, aside from not doing the reading at all, is to open the text to page 88 and begin reading. Reading an assignment is a three-phase process. It has a *before*, *during*, and *after* phase, and each is equally important!

The Before Phase

- Skim the entire assignment first so you have a general idea of the content, and try to relate the content to concepts you already know.
- Look at your syllabus and see how many lectures are devoted to the assignment. See what other material is included with this assignment on the next test.
- Judging from the length and the difficulty of the reading assignment, try to determine how long the task will take and plan ac-

cordingly. (Remember that reading technical information often takes more time than reading nontechnical works.)

The During Phase

- Mark up your text. Write notes in the margins. Highlight key concepts. (Try not to buy a book that has been marked up too much. You will be distracted by someone else's markings.)
- If examples do not show all the steps, write in the missing steps so you will remember how to do the problem when it is time to review for a quiz or test.
- Look at the pictures, figures, and graphs in the text. Remember, Alice in Wonderland asked, "What good is a book without pictures?" Fortunately, most modern chemistry books are filled with pictures. Take advantage of them. Remember that chemistry is visual. When the text refers to a drawing, figure, diagram, chart, photograph, etc., look at it right away. Study it. Sometimes one picture can be worth many words!
- Spend time reflecting on what you read. Can you put the concepts into your own words?
- Make a short outline of the material you have just read. Placing new concepts into your own words helps you to retain and master the new material.
- Do you need to hear yourself reading the material? Reading difficult material aloud slows you down and lets you hear as well as see. Two senses are better than one! Remember that you are reading for understanding.

The After Phase

- Go back to any sections you did not understand. If after going over them again, you still have concerns, put a question mark beside them and ask about them during the next class or help session or meeting with your study group.
- Go back over the entire reading assignment, looking at main headings and the notes you made in the margins. Does this information

remind you of the main ideas and content of each section? If not, add to it.

• Write a brief summary of what you read. Include any type of diagram that will help you remember the material. This summary, what you have underlined in the text, what you have written in the margins of the book, and the notes you have taken in class are the materials you need to use when you review daily and study for quizzes and tests.

14.3   Learning to Relax

It's hard to study effectively, and you certainly won't perform your best on an exam if your body is overly tense. Also, when you are tense, things always look worse than they are. However, some stress is normal when you enter into the unknown. And a little tension keeps you from getting too complacent and making careless errors. It is the excess tension that you need to be able to control.

There are three factors you can consciously control to reduce your stress and increase the effectiveness of your studying: posture, muscle tension, and breathing. We will discuss just a few techniques you can try with each factor.

Posture

When you are concentrating hard, especially when you are tired, you often tend to pull your neck down into your shoulder blades, collapse your chest and let the lower spine bow out. Sitting with the body in this position tends to make you tired and tense. Muscles work overtime to hold the body up, and the lungs cannot breathe efficiently.

It is important to have good posture while sitting and studying. If you cannot hold a good posture, you may need to take a break and relieve muscle tension.

Muscle Tension

A tense muscle uses more energy and requires more oxygen and nutrients than a muscle at rest. But a tense muscle constricts the capillaries that deliver oxygen and nutrients to it, which leads to fatigue of the muscle. The exhausted muscle becomes even more tense, starting a downward cycle. There are three things you can do to relieve muscle tension: stretching, massage, and movement.

* While you are studying, remember to stretch. Stand up periodically and stretch your body.
* Take a walk. Walking is one of the best ways to relax. Even a short walk around the block will help you relax and feel more alert.
* Periodically look out the window or across the room to relax your eyes.

Breathing

Your breathing is tied to your emotions. Have you noticed that after you have been scared, your breathing becomes rapid and shallow? This is one symptom of what is known as "fight or flight," a remnant of the days when we had to either confront a danger or run. In the modern world, this reaction does little to help us.

The same "fight or flight" reaction can happen on a smaller scale when you are studying unfamiliar material or when you are taking an exam. Here are some breathing techniques you can try anytime you feel tense.

* Breathe in slowly through your nose as you count to five.
* Then breathe out slowly.
* Pause for a second when you have completely exhaled, then repeat for a few more breaths.

When you inhale, expand your abdomen first to fill the lower lungs. Then let your chest rise to fill the upper lungs. Don't force the breath; let it come easy and natural. You can do this exercise a few minutes before you take an exam.

14.4   Conquering Tests

Preparing for the Test

- Test preparation begins on the first day of the course. Keeping current with all daily assignments is the most effective method of preparing for tests.
- Form a study group. You never really learn anything until you have to explain it to someone else. A study group will force you to do this. Working with others will also help you learn more study strategies. But study on your own both before and after the study group.
- Tests are more difficult than individual assignments because they contain a greater variety of material. To avoid getting confused, you must "overlearn" the material. To "overlearn," study actively until you really feel that you understand the material or can work the problems with no difficulty. Check to see how long you had to study or how many problems you had to work to reach this point. To "overlearn," you must spend at least half as much more time or work half as many more problems. Then you must also spend a small amount of time each day, maybe 10 to 15 minutes, reviewing past assignments after you have completed the new assignment.
- If there is a concept you do not understand, get help immediately. In chemistry, learning is often sequential, (i.e., new lessons build on past lessons). Therefore falling behind can be fatal!
- Try to predict test questions as you are studying.
- If practice tests are available, get them as soon as possible and do them at least several days before the test. Then you will still have time to get help if you find there are questions that are giving you difficulty.
- If you do fall behind in your studies, never try to learn new material the day or the night before a test. You will not have time to

"overlearn" it. Therefore, what will normally happen is you will confuse new material with what you already know. This confusion can cause you to miss even more test questions than you would have missed had you not tried to "cram in" the new material. Concentrate on what you know on the night before the test. Then, after the test is over, catch up immediately.

- The night before the test, be sure to get enough rest. Also check your calculator and put it and any other materials you will need in a place where you will be sure to pick them up before leaving for class.

Taking the Test

- *Wear a watch!* Bring your calculator, and make sure it is functioning.
- Get to class a few minutes early so you have time to get comfortable and get organized before the test is handed out.
- Put your name (and your student ID number) on the test!
- As soon as you get the test, write down on the test paper any information you are worried about forgetting. Then you won't have to worry anymore, and your mind will be free to concentrate completely on each test question.
- Preview the entire test. It only takes a minute or two but can tell you much. Previewing tells you the number and types of questions the test contains (multiple choice, true-false, matching, essay, problems, etc.), the number of points each question is worth, any additional information you need to write down to be sure you remember it, and how to schedule your time.
- Previewing should also build your confidence because if you are prepared, you will know immediately that you can answer most of the questions.
- Read any directions carefully and completely. If there is something you do not understand, raise your hand and ask about it. The instructor will probably be able to clarify it for you. In any case, the worst that can happen is the instructor will say he or she cannot answer your question.

- Begin with the questions that are easiest for you. This strategy will build your confidence and lessen your anxiety. It may also remind you of things you need to know to answer the more difficult items. In addition, it will prevent you from losing points by spending too much time on things you don't know and running out of time before completing items you do know.

- Return to questions you skipped. If they are objective questions, (i.e., true-false, multiple choice, matching, etc.), make an "educated guess." (See "Mastering the Multiple-Choice Test.") If there are essays or problems to be solved, do as much as you can. If the problem has several dependent parts and you do not know how to do the first part but do know how to do the next part, write a note saying, "I'm not sure how to do part A, so I am going to assume the answer is (make up a reasonable value) and use this value to complete the problem." Remember that partial credit is better than no credit! Answer every question as completely as possible.

- Be neat, and show your work; it will be easier to check your answers.

- Check your work! If you only have a few minutes left, check the first few questions you answered. You were most nervous then, and therefore you might have made careless errors. Then check the problems you thought were the easiest. You also make careless errors when you are too relaxed. If you eliminate careless errors, you can usually increase your grade by at least 10%!

- Do not get nervous because others are leaving before you. Those who finish early often are the weakest students. Use all the time allowed. There is no prize for finishing early!

Mastering the Multiple-Choice Test

In addition to the general test-taking strategies presented in this booklet, there are some specific things you can do to increase your score on multiple-choice tests.

- As soon as you get the test paper, jot down any information you know you will need and are afraid you will forget.

- Scan the entire test. If any of the questions remind you of any other facts you are afraid of forgetting, write them down.

- Read the directions. Most multiple-choice tests have only one correct choice, but don't assume anything.

- Answer the easiest, least time-consuming questions first. This strategy will assure that you have time to answer all the questions you know. It will also build your confidence and remind you of things that might help you answer the more difficult items.

- Before you look at the answer choices, work the problem and ask yourself, "Does this answer make sense?" Remember that many of the incorrect answer choices will be derived from mistakes that students commonly make. If you see your answer as one of the choices, don't immediately mark it.

- If you don't find your answer among the choices, look again. Perhaps your answer is there but in a different form. For example, 1/2 could be written as 0.5, 0.50, or 50%, etc. Maybe the answer is in scientific notation, and you calculated your answer in decimal form.

- After you've answered all the questions you know, carefully re-read each question you skipped. You may find you now remember how to do them. Doing the questions you know warms you up, and sometimes answering one question will spark your memory on how to answer another.

- If you are still unable to answer some of the questions, make an "educated" guess. Can you eliminate one or more of the choices? For example, if the question asks for grams, eliminate all the choices that don't have units of grams.

- Never leave a multiple-choice question blank unless the directions say there is a penalty for guessing.

- Use all the time allowed. There is no prize for being the first to finish.

Take time to read this advice and integrate these ideas into your study habits and test-taking strategies. I have seen them work for many students. They can work for you and aid you in your mastery of Introductory Chemistry.